Special English for the Students of Biology

Dr. Azadeh Nemati

Department of English Language Teaching, Jahrom Branch, Islamic Azad University, Jahrom, Iran.

Copyright © 2013 Islamic Azad University, Jahrom, Iran.
All rights reserved.
ISBN: 978-964-10-1511-6

ABOUT THE AUTHOR

Dr. Azadeh Nemati is an assistant Professor in Iran, majoring in ELT. She is the editor in chief of some international journals and has already published books and articles nationally and internationally. She has also supervised some MA theses. In 2010 and 2012 she was selected as distinguished researcher in the University.

Contents of the book

Unit 1: Biologically speaking who are we? 1
Unit 2: Cell structure and function 6
Unit 3: Agents and Their Important Features 12
Unit 4: Eukaryotes and prokaryotes 18
Unit 5: Animals versus Plants 24
Unit 6: What is a plant? 30
Unit 7: Human chromosomes 36

Complimentary texts for reading and translation 42
Text 1: The animal cell 43
Text 2: Cell division 47
Text 3: Physical properties 49
Text 4: Kingdoms of organisms 51

References 54
Appendix: Common root words 55
Glossary 66

The International Phonetic Alphabet (IPA)

The International Phonetic Alphabet is an alphabetically system of phonetic notation based primarily on the Latin alphabet. It was devised by the international phonetic association as a standardized representation of the sounds of spoken language. It provides the academic community world-wide with a notational standard for the phonetic. Foreign language learners can use it to improve their pronunciation while speaking.

vowels	
IPA	**Examples**
ʌ	c*u*p, l*u*ck
ɑ:	*a*rm, f*a*ther
æ	c*a*t, bl*a*ck
e	m*e*t, b*e*d
ə	*a*way, cinem*a*
3:ʳ	t*ur*n, l*ear*n
ɪ	h*i*t, s*i*tt*i*ng
i:	s*ee*, h*ea*t
ɒ	h*o*t, r*o*ck
ɔ:	c*a*ll, f*ou*r
ʊ	p*u*t, c*ou*ld
u:	bl*ue*, f*oo*d

Diphtongue	
aɪ	f*i*ve, *eye*
aʊ	n*ow*, *ou*t
eɪ	s*ay*, *eigh*t
oʊ	g*o*, h*o*me
ɔɪ	b*oy*, j*oi*n
eəʳ	wh*ere*, *air*
ɪəʳ	n*ear*, h*ere*
ʊəʳ	p*ure*, t*ou*rist

Consonants	
IPA	Examples
b	*b*ad, la*b*
d	*d*id, la*d*y
f	*f*ind, i*f*
g	*g*ive, fla*g*
h	*h*ow, *h*ello
j	*y*es, *y*ellow
k	*c*at, ba*ck*
l	*l*eg, *l*ittle
m	*m*an, le*m*on
n	*n*o, te*n*
ŋ	si*ng*, fi*n*ger
p	*p*et, ma*p*
r	*r*ed, t*r*y
s	*s*un, mi*ss*
ʃ	*sh*e, cra*sh*
t	*t*ea, ge*tt*ing
tʃ	*ch*eck, *ch*ur*ch*
θ	*th*ink, bo*th*
ð	*th*is, mo*th*er
v	*v*oice, fi*v*e
w	*w*et, *w*indo*w*
z	*z*oo, la*z*y
ʒ	plea*s*ure, vi*s*ion
dʒ	*j*ust, lar*ge*

Unit 1

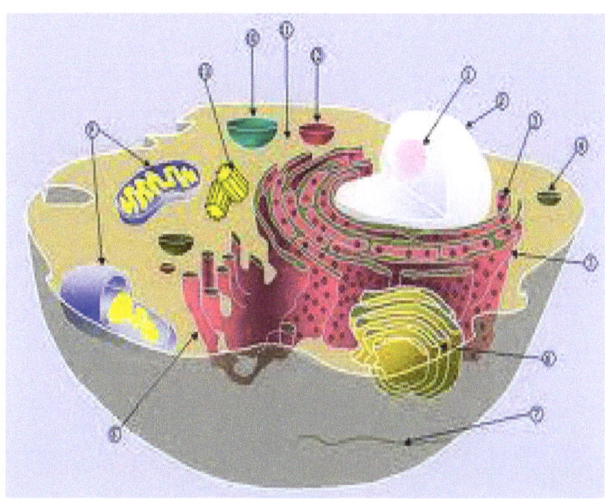

A) New vocabularies (before you read the text)

Before you read the text try to find the pronunciation and the meaning of the following words using IPA symbols in a dictionary. Fill in the blanks and use the new vocabulary items in a sentence.

Launch /……………...../ (   ) Example:……………………….

Appropriate/…………..../ (   ) Example:……………………….

Characteristics/………../ (   ) Example………………………..

Require/……………….../(   ) Example………………………..

Digestive/…………….../(   ) Example………………………..

Continuance/…………./(   ) Example………………………..

Evolution/ …………...../(   ) Example………………………..

Diversity/……………../(   ) Example………………………..

Biologically speaking who are we?

If you are about to **launch** on the study of human biology, before beginning it is **appropriate** to define who humans are and how they fit into the world of living.

Certain **characteristics** tell us who human beings are biologically speaking.

Human beings are highly organized. A cell is the basic unit of life, and human beings are mutlicellular since they are composed of many types of cell. Like cells from tissues, and tissues make up organs. Each type of organ is part of an organ system. The different systems perform the specific functions. Together, the organ systems maintain homeostasis, an internal environment for cells that varies only within certain limits. For example, cells **require** a constant supply of nutrients and they give off waste products. The **digestive** system takes in nutrients, and the circulatory system distributes these to the cells. The waste products are excreted by the excretory system. The work of the nervous and endocrine systems is critical because they coordinate the functions of the other system.

Human beings reproduce and grow. Reproduction and grow are fundamental **characteristics** of all living things. Just as cells come only from pre-existing cells, so living things have parents. When living things reproduce, they create a copy of themselves and assure the **continuance** of the species. (A species is a type of living things). Growth occurs as the resulting cell develops into the newborn. Development includes all changes that occurs from the fertilized egg to death and, therefore, all the changes that occur during childhood, adolescence, and adulthood.

Humans have a cultural heritage. We are born without knowledge of civilized ways of behaving , and we gradually acquire these by adult instruction and imitation of role models. It is our cultural inheritance that makes us think we are separate from nature. But actually we are a product of **evolution,** a process of change that has resulted in the **diversity** of life, and we are a part of the biosphere, a network of life that spans the surface of earth.

B) Comprehension questions (after you read the text)

Answer the following questions.

1) Describe some characteristics that express biologically speaking who human beings are.

....

2) What is the diference between multicellular and unicellular creatures?

....

3) What is development?

....

4) Compare and contrast childhood, adolescence, and adulthood.

....

C) Match the following system to its related function.

System	function
1) Digestive	a) supports and moves organism
2) Circulatory	b) transports nutrients to and waste from cells
3) Immune	c) produces offspring
4) Respiratory	d) convert food particles to nutrient molecules
5) Excretory	e) exchange gases within the environment
6) Nervous	f) eliminates metabolic wastes
7) Musculoskeletal	g) regulates systems and internal environment
8) Endocrine	h) defends against diseases
9) reproductive	i) regulates systems and internal environment

D) Complete the following sentences from the text.

1) A cell is ....

2) Growth occurs ....

3) Development includes ....

4) Evolution can be defined as ....

5) Nervous and endocrine systems ....

E) Cloze test

Fill in the blanks with the appropriate words given bellow.

History   cells   presence   trace   bacteria   product
ancestors   genetic

Human beings are a………….. of an evolutionary process. Life has a ……….that began with the evolution of the……….about 3.5 billion years ago. It is possible to ……human ancestry from the first cell through a series of prehistoric…………..until the evolution of modern-day humans. The ……….of the same type of chemicals tell us that human beings are related to all other things. DNA is the ……….material, and ATP is the energy currency in all cells, including human cells. It is even possible to do research with ………..and have the results apply to humans.

Scope of microbiology

Microbiology is the study of microbes, organisms so small that a micrscope is needed to study them. Two dimensions of the scope of microbiology are considered namely, 1) the variety of kinds of microbes and 2) the kinds of work microbiologist do.

The major groups of organisms studied in microbiology are bacteria, algae, fungi, viruses, and protozoa. All are widely distributed in nature. For example, a recent study of bee bread (a pollen-driven nutrients eaten by worker bees) showed it to contain 188 kinds of fungi and 29 kinds of bacteria. Most microbes consist of a single cell. Cells are the basic units of structure and function in living things.

Viruses, tiny cellular living and nonliving, behave like living organisms when they gain
entry to cells. They, too are studied in microbiology. Microbes range in size from small viruses 20 nm in diameter to large protozoans 5 mm or more in diameter. In other words, the largest microbes are as much as 250,000 times the size of the smallest ones!

Each unit of this book, in the part of "take a look" will give a short explanation of **bacteria** (unit 2), **algae** (unit 3), **fungi** (unit 4), **viruses** (unit 5), **protozoa** (unit 6), and finally macroscopic **helminthes** (unit 7) respectively.

Unit 2

A) New vocabularies (before you read the text)

Before you read the text try to find the pronunciation and the meaning of the following words using IPA symbols in a dictionary. Fill in the blanks and use the new vocabulary items in a sentence.

Sacrificing /……….…..../ (   ) Example:………………………….

Snooz/……………...…../ (   ) Example:…………………………..

Pace/…………..………../ (   ) Example…………………………..

Fermentation /……….…/(   ) Example…………………………..

Fatigue /………………../(   ) Example……………………………..

Fundamental /…………./(   ) Example……………………………

Imagine / ……………..../(   ) Example……………………………..

Cell structure and function

Sacrificing his usual Saturday morning **snooze**, Michael decides to join friends in a nearby 30-mile cycling event. He was an experienced cyclist, but things were going well-until about halfway through the ride. Then, grinding his way up a hill, Michael simply ran out of steam. His legs felt rubbery. He gasped for air. He got off the bike and sat down.

What happened? Michael had forgotten to **pace** himself. His leg muscles worked so hard they used up oxygen faster than it could be resupplied. Without oxygen, cells switch to a kind of energy production called **fermentation**. One unwanted by-product of fermentation is lactic acid, which begins to collect in skeletal muscles and makes muscle contraction more difficult.

Eventually, lactic acid builds to the point that it causes muscles **fatigue**. In cycling lingo, this is called the dreaded "bonk". To recover, an athlete must slow down and breathe heavily; giving badly needed oxygen back to the body.

Our bodies are composed of many cells, and although each type is specialized for a particular function, all cells have certain basic structure and metabolism. The cell is the **fundamental** unit of our bodies, and it is at the cellular level that we must understand how the body functions. Michael's fatigue began with each individual muscle cell and not with the muscle as a whole. Because cells are microscopic, it is sometimes hard to imagine that individual muscle cells account for the functioning of an organ like a skeletal muscle.

B) Comprehension questions (after you read the text)

Put T if the sentence is right and F if it is false.

1) Michael decides to join friends in skiing.

2) Michael's muscles used up oxygen slower than it could be resupplied.

3) Lactic acid causes muscle fatigue.

4) Since all cells are specialized their metabolism is different.

5) The cell is the basic unit of body.

C) Two/ three part phrasal verbs

A two-part verb consists of a simple verb and one or more short words which are called particles. Very often, the meaning of a two-part verb seems to have nothing to do with the meaning of the main verb or the meaning of the particle. For example consider the following sentence

Jill and David had a big fight, but then they <u>made up.</u>

This sentence means that after a big fight, Jill and David had a talk and then were no longer angry with each other. They did not make anything and nothing went up!

English has hundreds of two-part verbs. Each time you learn one of these, you should learn it as a single unit. Now try to complete the following text about a teacher talking to her/his students. The verbs are given below:

take out   turn off   put away   hang up   clean up   take off   turn on

Ok class! Time to start. Mary could you……………the lights? Thanks. OK everyone,………… your books,……………….. and make sure you…………………your cell phones. Hey! No cell phones in class!………….that………………………!

Actually, before we start I want to tell you a little story. Every night I go home……………..my suit and throw it on the bed. My wife always says "You'd better……………….your suit or it will get wrinkled. Of course I do what she says, but last night she was really angry. She said "You're so messy. I always have to.................................after you.

Now please refer to the text and find some more two or three phrasal verbs such as, ran out of, got off and …. and write their meanings.

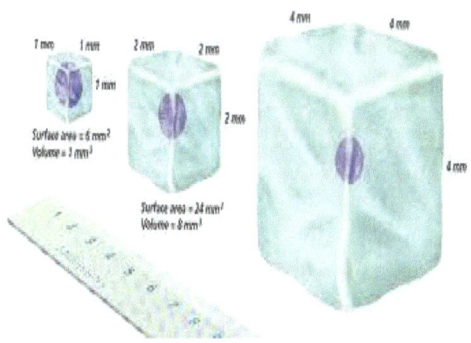

Cell sizes

D) Cloze test

Fill in the blanks with the appropriate words given bellow.

Functions   obtaining   composed   naked eye   invention
tiny

All living things are made up of ………… fundamental units called cells. Because cells are so small, the study of cells did not begin until the …………..of the first microscope in the seventieth century. Then the cell theory, which states that all living things are …………of cells, and new cells arise only from preexisting cells, was formulated. Regardless of cells size and shape, it must carry on the …………..associated with life; interacting with the environment,………… chemicals and energy, growing and reproducing. A few cells like a hen's egg or a frog's egg are large enough to be seen by the ……………… but most are not.

E) Brain Teaser

Do you know why cells stay small?

Because a cell needs a surface area that can adequately exchange materials with the

environment, this explains why cells stay small.

Bacteria

Among the great variety of microorganisms that have been identified, bacteria probably have been the most thoroughly studied. The majority of bacteria (singular: bacterium) are single –celled organisms with spherical, rod, or spiral shapes, but a few types form filaments. Most are so small they can be seen with a light microscope only under the highest magnification. Although bacteria are cellular, they don't have a cell nucleus, and they lack the membrane-enclosed intracellular structures found in most other cells.

Many bacteria absorb nutrients from their environment, but some make their own nutrients by photosynthesis or other synthesis processes. Some are stationary, and others move about. Bacteria are widely distributed in nature, for example in aquatic environments and in decaying matter. And some occasionally causes disease.

Unit 3

A) New vocabularies (before you read the text)

Before you read the text try to find the pronunciation and the meaning of the following words using IPA symbols in a dictionary. Fill in the blanks and use the new vocabulary items in a sentence.

Agent /…………………/ (   ) Example:…………………………

Organisms /……………/ (   ) Example:…………………………

Kingdom /……………. / (   ) Example…………………………..

Multicellular /…………. …/(   ) Example…………………………..

Salient /………………../(   ) Example…………………………….

Surrounded /……….…../(   ) Example…………………………….

Replicate / …………….../(   ) Example……………………………..

Agents and Their Important Features

The **agents** of human infectious disease belong to 5 major groups of **organisms**: bacteria, fungi, protozoa, helminthes and viruses. The first 3 groups are members of the **kingdom** of protists, one of the primary biologic subdivisions along with animals and plants. The protists are distinguished from animals and plants by being either unicellular or relatively simple **multicellular** organisms that are classified as metazoan within the animal kingdom. Taken together, the helminthes, and the protozoa are commonly called parasites. Viruses are quite distinct from organisms, as they ar not cells but can replicate only within cells.

One **salient** feature is that bacteria, fungi, protozoa, and helminthes are cellular, whereas viruses are not. This distinction is based primarily on 3 criteria:

1) Structure. Cells have a nucleus or nucleoid containing DNA; this is **surrounded** by cytoplasm. Within which proteins are synthesized and energy is generated. Viruses have an inner core of genetic material but no cytoplasm, and so they depend on host cells to provide the machinery for protein synthesis and energy generation.

2) Method of replication. Cells **replicate** by binary fission, during which one parent cell divides to make 2 progeny cells while retaining its cellular structure. In contrast, viruses disassemble, produce many copies of their nucleic acid and protein, and then reassemble into multiple progeny viruses. Furthermore, viruses must replicate within host cells, because, as mentioned above they lack protein-synthesizing and energy-generating system. With the exception of rickettsiae and chlamydiae, which are bacteria that also require living host cells for growth, bacteria can replicate extra cellularly.

3) Nature of the nucleic acid. Cells contain both DNA and RNA, whereas viruses contain either DNA or RNA but not both.

B) Comprehension questions (after you read the text)

Choose the best answer.

1) The ………….. can be distinguished from animals and plants by being unicellular or multicellular.

a) bacteria   b) fungi   c) viruses   d) protozoa

2) Being …………...is the characteristics of bacteria, fungi, protozoa and helminthes.

a) multicellular   b) cellular   c) non- cellular   d) unicellular

3) Viruses lack which one ?

a) inner core   b) cytoplasm   c) protein-synthesizing   d) both b and c

4) Viruses contain ……………

a) DNA   b) RNA   c) DNA or RNA   d) DNA and RNA

C) Based on the text try to complete the following table

Kingdom	Pathogenic microorganisms
Animal	
	None
	Fungi

D) Prefixes and suffixes

Most English words used today were not originally English. These words were adapted from other language, such as Latin and Greek. It may be possible to guess the meaning of an unknown word when one knows the meaning of its root. Knowing prefixes and suffixes can also help.

An English word can consist of three parts: the root, a prefix, and a suffix. The **root** is the part of the word that contains the basic meaning, or definition of the word. The **prefix** is a word element placed in front of the root, which changes the word's meaning or makes a new

Common prefixes

Prefix	Meaning	Example
pre-	before	They will show a sneak *preview* of the movie.
un-	not	The cafeteria will be *unavailable* tomorrow morning.
dis-	not	Mark *disagreed* with John's philosophy.
re-	again	Are you going to *renew* your subscription?
mis-	not	He has *mismanaged* the company.
im-	not	With hard work and determination, nothing is *impossible*.
bi-	two	Henry recently received his first pair of *bifocals*.
de-	not	Many ecologists are concerned about the *deforestation* of our world's rain forests.

Common suffixes

Suffix	Meaning	Example
-er	doer	I work as a computer *programmer*.
-able	able	These glass bottles are *recyclable*.
-ous	full of	Driving on the freeway can be *dangerous*.
-ness	state of being	At night, the earth is covered in *darkness*.
-ful	full of	The witness gave an honest and *truthful* testimony.
-ly or -y	like	James whistled *happily* on his way home from school.
-ment	state of	Mary sighed with *contentment*.

The list of some more affixes (prefix and suffix) are included in Appendix one.

Now please read the text again and fine more vocabularies with affix and suffix and try to guess the meaning of them.

Algae

In contrast to bacteria which were studied in the previous unit, several groups of microorganisms consist of larger, more complex cells that have a cell nucleus. They include algae, fungi, and protozoa, all of which can easily be seen with a light microscope. The first one ,i.e., algae, will be describe in this unit

Many algae (singular, alga) are single-celled microscopic organisms, but some marine algae are large, relatively complex, multicellular organisms. Unlike bacteria, algae have a clearly defined cell nucleus and numerous membrane-enclosed intracellular structures. All algae photosynthesize their own food as plants do and many can move about.

Algae are widely distributed in both fresh water and oceans. Because they are so numerous and because they capture energy from sunlight in the food they make, algae are an important source of food for other organisms. Algae are of little medical importance; only one species, prototheca, has been found to cause disease in humans. Having lost its chlorophyll, and therefore the ability to produce its own food, it now makes meals of human.

Unit 4

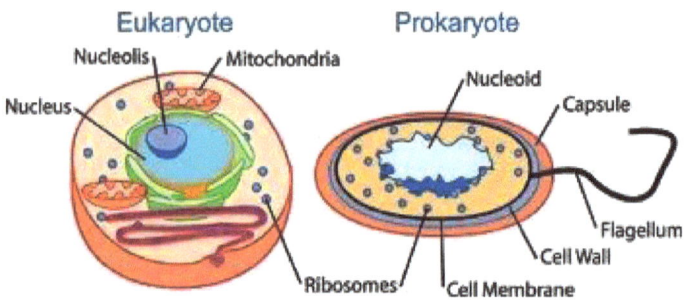

A) New vocabularies (before you read the text)

Before you read the text try to find the pronunciation and the meaning of the following words using IPA symbols in a dictionary. Fill in the blanks and use the new vocabulary items in a sentence.

Evolve /…………………/ (   )   Example:……………………………

Apparatus /……………./ (   )   Example:…………………………..

Lack /…..………………/ (   )   Example……………………………

Rigid /…………………..../(   )   Example……………………………

Motility /……………....../(   )   Example……………………………..

Possess /………....…..../(   )   Example……………………………..

Organ / ……….…..……./(   )   Example……………………….

Eukaryotes and prokaryotes

Cells have **evolved** into two fundamentally different types, eukaryotes and prokaryotes, which can be distinguished on the basis of their structure and the complexity of their organization. Fungi and protozoa are eukaryotic, whereas bacteria are prokaryotes.

1) The eukaryotes cell has a true nucleus with multiple chromosomes surrounded by nuclear memberane and uses a mitotic **apparatus** to ensure allocation of the chromosomes to progeny cells.

2) The nucleoid of a prokaryotic cell consists of a single circular molecule of loosely organized DNA **lacking** a nuclear membrane and mitotic apparatus.

In addition to the different types of nuclei, the 2 classes of cells are distinguished by several other criteria.

First, eukaryotic cells contain organelles, such as mitochondria and lysosomes, and larger (80S) ribosomes, whereas prokaryotes contain co organelles and smaller (70S) ribosomes.

Second, most prokaryotes have **rigid** external cell wall that contains peptidoglycan, a polymer of amino acids and sugars, as its unique structural component. Eukaryotes, on the other hand, do contain peptidoglycan. Either they are bound by a flexible cell membrane or, in the case of fungi, they have a rigid cell wall within, a homopolymer of N-acetylglucocamine, typically forming the framework.

And finally, the eukaryotic cell membrane contains sterols, whereas no prokaryotes, except the wall-less Mycoplasma, has sterols in its membrane.

Another criterion by which these organisms can be contrasted is **motility**. Most protozoa and some bacteria are motile, whereas fungi and viruses are non motile. The protozoa are a heterogeneous group that **possesses** 3 different **organs** of locomotion: flagella, cilia, and pseudopods.

The motile bacteria move only by means of flagella.

B) Comprehension questions (after you read the text)

Answer the following questions.

1) What are the differences between bacteria and viruses?
....
2) Are bacteria prokaryotic or eukaryotic?
....
3) What are the differences between prokaryotic and eukaryotic cells?
....
4) What are the similarities and differences among bacteria, fungi and protozoa?
....

C) Read the text again carefully and complete the table with yes or no.

characteristics	Prokaryotic bacterial cells	Eukaryotic human cells
DNA within a nuclear membrane		
Mitotic division		
DNA associated with histones		
Chromosome number	One	More than one
Membrane-bound organelles such as mitochondria and lysosomes		
Size of ribosome	70S	80S
Cell wall containing peptidoglycan		

D) Word study

Adjective	Adverb	Noun	Verb
Rigid	Rigidly	Rigidity	....
Distinguished/distinguishable Distinguish	distinguishably	....	
Organized/organizational organize	organizationally	organization/organizer	

Try to complete the sentences using the above words.

1) The curriculum was too............ .

2) The speed limit must be ………….. enforced.

3) He sat upright, her body ………….with fear.

4) At what age children are able to …………..between the right and wrong?

5) The male bird is easily…………..from female.

6) He is the president of a large international ……………... .

7) We don't fully understand how the brain is …………….. .

8) To register the seminar you must see the…………….... .

9) If you like to improve you must have a well-………… office.

E) Plural form of noun

The plural form of most nouns is created simply by adding the letter *s*.

- more than one snake = snakes
- more than one ski = skis

Words that end in *-ch, x, s* or *s-like* sounds, however, will require an *-es* for the plural.

- more than one witch = witches
- more than one box = boxes
- more than one gas = gases
- more than one bus = buses
- more than one kiss = kisses
- more than one Jones = Joneses

There are several nouns that have irregular plural forms. Plurals formed in this way are sometimes called **mutated (or mutating) plurals**.

- more than one child = children
- more than one woman = women
- more than one man = men
- more than one person = people
- more than one goose = geese
- more than one mouse = mice
- more than one barracks = barracks
- more than one deer = deer

And, finally, there are nouns that maintain their Latin or Greek form in the plural.

- more than one nucleus = nuclei
- more than one syllabus = syllabi
- more than one focus = foci
- more than one fungus = fungi
- more than one cactus = cacti (*cactuses* is acceptable)
- more than one thesis = theses
- more than one crisis = crises
- more than one phenomenon = phenomena
- more than one index = indices (*indexes* is acceptable)
- more than one appendix = appendices (*appendixes* is acceptable)
- more than one criterion = criteria

Fungi

Like algae, many fungi (singular, fungus) such as yeast and some molds are single-celled microscopic organisms. Some such as mushrooms are multicellular, macroscopic organisms. Fungi also have a cell nucleolus and intracellular structure. All fungi absorb ready- made nutrients from their environment. Some fungi form extensive networks of branching filaments, but the organisms themselves generally do not move. Fungi are widely distributed in water and soil as decomposers of dead organisms. Some are important in medicine either as agents or diseases such as ringworm and vaginal yeast infections or as sources of antibiotics.

Unit 5

A) New vocabularies (before you read the text)

Before you read the text try to find the pronunciation and the meaning of the following words using IPA symbols in a dictionary. Fill in the blanks and use the new vocabulary items in a sentence.

Zoology /……………..…/ (   ) Example:………………………….

Includes /……………....../ (   ) Example:…………………………..

Properties /….....…......./ (   ) Example…………………………..

Invariable /…………….../(   ) Example…………………………..

Pigment /……………........./(   ) Example…………………………..

Release /……….....…..../(   ) Example……………………………

Stimuli / ……….…….…/(   ) Example…………………………...

Animals versus Plants

The science dealing with animal life is **zoology** (Greek Zoon, animal + logos, discourse). It **includes** all common knowledge about animals together with more technical data. Zoology deals with the structure and bodily functions of animals, their habits, where and how they live, their relations with one another and with their environments their classification, and many other features. In short, all the facts, conclusions, theories, and laws relating to animal life make up the science of zoology. Together with botany, the science of plants, it forms biology (Greek, bio, life), the science of life. These and other fields dealing with the phenomena of nature such as geology (earth structure), physiography (earth-sruface feature), meteorology (atmosphere), and so forth are the natural sciences. They stand in contrast to the physical sciences- physic, the **properties** of matter, and chemistry, the composition of matter.

Some important differences between plants and animals follow:

1) Form and structure. The animal body form is rather **invariable**, the organs are mostly internal, growth usually produces changes in proportions with age, the cell membranes are delicate and the body fluids contain sodium chloride (NaCl). In plants the form often is variable, organs are added externally, the cells commonly are within thick cellulose walls, and sodium chloride usually is toxic. Most growth is at the ends of organs and often continues throughout life. But each kind of plant has a normal growth limit.

2) Metabolism. Animals require complex organic materials as food, obtained by eating plants or other animals. These foods are broken down an reorganized chemically within the body. Oxygen (O2) is usually needed for respiration. The end products of metabolism are mainly carbon dioxide (CO2), water, carbon dioxide from the air, and inorganic chemicals obtained in solution from soil. By photosynthesis_ the action of sunlight on the green **pigment** known as chlorophyll_ these simple materials are formed into

various organic compounds and oxygen **release** as a by product

3) Nervous system and movement. Most animals have a nervous system and respond quickly to **stimuli**; plants have no such systems and react slowly. Animals commonly can move about or move parts of the body but certain kinds became fixed early in life and other fixed forms are plant-like form.

B) Comprehension questions (after you read the text)

Answer the following questions.

1) Describe the science of zoology.

....

2) Name major differences between animals and plants.

....

3) Explain metabolism in your own words.

....

4) Write about nervous system of animals.

....

C) Like other sciences, zoology has been subdivided as a result of great growth of Knowles. Write the function of each branch below.

Cytology………………………………………………………..

Morphology……………………………………………………

Histology………………………………………………………..

Physiology……………………………………………………

Embryology ……………………………………………………

Genetics………………………………………………………..

Behavior………………………………………………………

Ecology………………………………………………

Zoogeography………………………………………

Paleonthology………………………………………

Evolution……………………………………………

Taxonomy……………………………………………

Protozoology………………………………………..

Entomology ………………………………………...

D) Do they know any idioms in English which contain animals?

     Idioms are words, phrases, or expressions that cannot be taken literally. In other words, when used in everyday language, they have a meaning other than the basic one you would find in the dictionary. Every language has its own idioms. Learning them makes understanding and using a language a lot easier and more fun!

Try to learn these animal idioms.

1. Like a dog with two tails - to be extremely happy
2. A little bird told you - someone told you a secret
3. A leopard can't change his/her spots - people can't change themselves.
4. To be no spring chicken - to be relatively old
5. Copycat - to do the same thing as someone else

6. The cat has got his/her tongue - to be too nervous to speak
7. To be fishy - to be strange/wrong; possibly criminally wrong
8. To have butterflies - to feel nervous
9. Donkey's years - a very long time
10. To eat a horse - to be able to eat a lot
11. To get someone's goat - to really irritate him/her
12. To look a gift horse in the mouth - to not be grateful for something
13. To talk the hind leg off a donkey - to talk a lot for a long time
14. To drink like a fish - to drink a lot
15. To be in the doghouse - to be in trouble with someone/ for someone to be angry with you
16. To horse around - to play around/joke
17. To be as quiet as mice - to be silent
18. To make a mountain out of a molehill - to exaggerate a problem
19. Not enough room to swing a cat - a place is very small
20. To grin like a Cheshire cat - to have a very big smile.

Viruses

Viruses are acellular entities too small to be seen with a light microscope. They are composed of specific chemical substances- a nuclei c acid and a few proteins. Indeed, some viruses can be crystallized and stored in a container on a shelf for years, but they retain the ability to invade other cells. Viruses replicate themselves and display other properties of living organisms only when they have invaded cells. Many viruses can invade human cells and cause disease. Even smaller acellular agents of disease are viroids (nucleic acid without a protein coating) and prions (proteins without any nucleic acid). Viriods have been shown to cause various plants disease, whereas prions cause mad cow disease and related disorders.

Unit 6

A) New vocabularies (before you read the text)

Before you read the text try to find the pronunciation and the meaning of the following words using IPA symbols in a dictionary. Fill in the blanks and use the new vocabulary items in a sentence.

Provide /…… …….…../ (   ) Example:………………………….

Overwhelming /……..../ (   ) Example:…………………………..

Recognize /…..……..../ (   ) Example……………………………..

Heterogeneous /…….…..../(   ) Example…………………………….

Affinity /……………......../(   ) Example…………………………..

Obvious /…………......……/(   ) Example……………………………

Convenience / …….…..../(   ) Example……………………………..

What is a plant?

The science of plant biology is primarily the study of flowering plants or angiosperms. Flowering plants are by far the most important group of plants in the world, **providing** the **overwhelming** majority of plant species and most of the biomass on land. They are the basis for nearly all our food.

Historically, the science of plant biology, or botany, has included all living organism except animals, but it is clear that there is a major division of life between cells with a simple level of organization, the prokaryotes and those with much more complex cells the eukaryotes. The prokaryotes include bacteria and bacteria-like organisms. Among eukaryotes three main multicellular kingdoms are **recognized**: animals, plants, and fungi. There is a fourth **heterogeneous** group of eukaryotes that are mainly unicellular but with a few multicellular groups such as slime molds and large algae. Some of these have **affinity** with animals, some with plants, some with fungi and some have no **obvious** affinity. They are grouped together, for **convenience**, as a kingdom, the protists. There is no clear boundary between protists and plants and authors differ in which organisms they consider in which kingdom.

Plants are photosynthetic and autotrophic, have chlorophyll a and b except for some algae, have a cellulose cell wall and vacuole, and have an alternation of diploid and haploid generation. Vegetative structure is similar across vascular plants; reproductive structures differ.

B) Comprehension questions (after you read the text)

Answer the following questions.

1) What is plant biology?
....
2) Are plants eukaryotic or prokaryotic? Explain.
....
3) Write features of plants.
....

C) Try to draw different parts of a typical plant cell in the following cell wall

c) Complete the following sentences.

There are some characteristics that define plants as different from other eukaryotes such as:

1) Plants are photosynthetic, it means...

2) The photosynthetic pigment is chlorophyll, and it is ....

3) The cells have a cell was made of...

4) There is an alternation of diploid and haploid, in other words................

Idioms from Plants

Each example below has an idiom that contains a word related to plants. Can you guess the meaning of each idiom from the context? Try to match each idiom (1-6) with its definition (a-f).

- He's crazy about Hong Kong so he tends to see everything in Hong Kong *through rose-tinted glasses*.
- Raising two children on one salary is *no bed of roses*.
- After taking a long, hot shower, he felt *fresh as a daisy*.
- What? 20,000 yen for a shirt? *Money doesn't grow on trees*, you know. Be more economical!
- I saw her just before her talk and she was *shaking like a leaf*.
- He has *turned over a new leaf* and he's not drinking any more.
-

Idiom	Definition
1. to see through rose-tinted glasses	a. to start behaving in a better way
2. no bed of roses	b. to shake a lot because of fright or nervousness
3. to be fresh as a daisy	c. to see only the pleasant parts of something
4. money doesn't grow on trees	d. a situation that is difficult or unpleasant
5. to shake like a leaf	e. to be full of energy and enthusiasm
6. to turn over a new leaf	f. money is not easy to get

Protozoa

Protozoa (singular, protozoan) also are single-celled, microscopic organisms with at least one nucleus and numerous intracellular structures. A few species of amoebae are large enough to be seen with the naked eye, but we can study their structure only with a microscope. Many protozoa obtain food by engulfing or ingesting smaller microorganisms. Most protozoa can move, but a few, especially those that cause human disease cannot. Protozoa are found in a variety of water and soil environments, as well as in animals such as malaria-carrying mosquito.

Unit 7

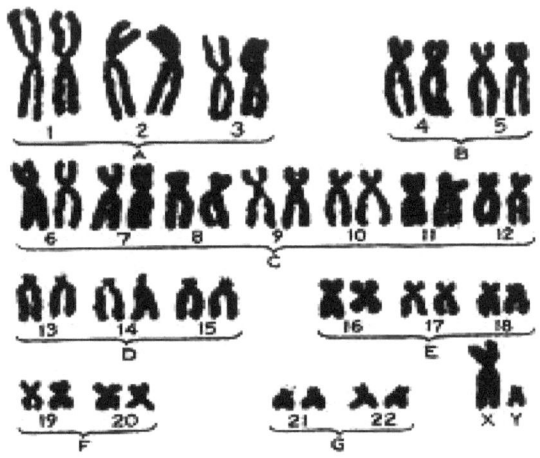

A) New vocabularies (before you read the text)

Before you read the text try to find the pronunciation and the meaning of the following words using IPA symbols in a dictionary. Fill in the blanks and use the new vocabulary items in a sentence.

Division /……………../ (   ) Example:…………………….

Somatic /……………./ (   ) Example:…………………….

Regeneration /…………/ (   ) Example……………………..

Identical /………………/(   ) Example……………………..

Occur /………………./(   ) Example……………………..

Abnormalities /………../(   ) Example……………………..

Arise / ..../( ) Example...........................

Human chromosomes

There are two kinds of cell **division**: mitosis and meiosis. Mitosis is ordinary **somatic** cell division, by which the body grows, differentiates, and effects tissue **regeneration**.

Mitotic division normally results in two daughter cell, each with chromosomes and genes **identical** to those of parent cell. There may be dozens or even hundreds of successive mitoses in a lineage of somatic cell.

In contrast, meiosis **occurs** only in cells of the germline. Meioses results in the formation of reproductive cells (gametes) each of which has only 23 chromosomes: one of each kind of autosome and either an X or a Y. Thus, whereas somatic cells have the diploid (double) or the 2n chromosome complement (i.e., 46 chromosomes), gamets have the haploid (single) or the n complement (i.e., 23 chromosome). **Abnormalities** of chromosome number or structure, which are usually clinically significant, can **arise** in either somatic cell or cells of the germline by errors in cell division.

B) Comprehension questions (after you read the text)
Answer the following questions.
1) What do you know about cell division?
....
2) What is daughter cell?
....
3) Compare and contrast mitosis and meiosis.
....
4) How many chromosomes human beings have?
....

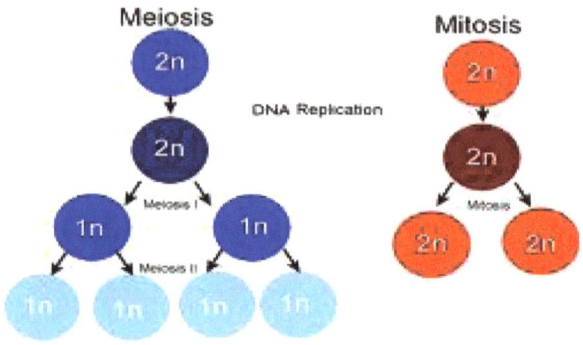

C) In your own language explain meiosis and mitosis again. You can use the above diagram as well.

....
....
....
....
....
....
....

D) Cloze test

Fill in the blanks with the appropriate words given bellow.

Critical   appropriate   determinants   histones   nucleus   naked   chromosomes

Structure of human chromosomes

The composition of genes in human genomes, as well as the............ of their expression, is specified in the DNA of the 46 human........... . Each human chromosomes is believed to consist of a single, continuous DNA double helix; that is each chromosome in the ....is a long, linear, double-stranded DNA molecule. Chromosomes are not ....DNA double helices however. The DNA molecule of a chromosome exits a s a complex with a family of basic chromosomal proteins called ....and with a heterogeneous group of acidic, nonhistone ....that are much less well characterized, but that appear to be ....for establishing a proper environment to ensure normal chromosome behavior and ....gene expression. Together, this complex of DNA and protein is called...................... .

E) Translate the following passage

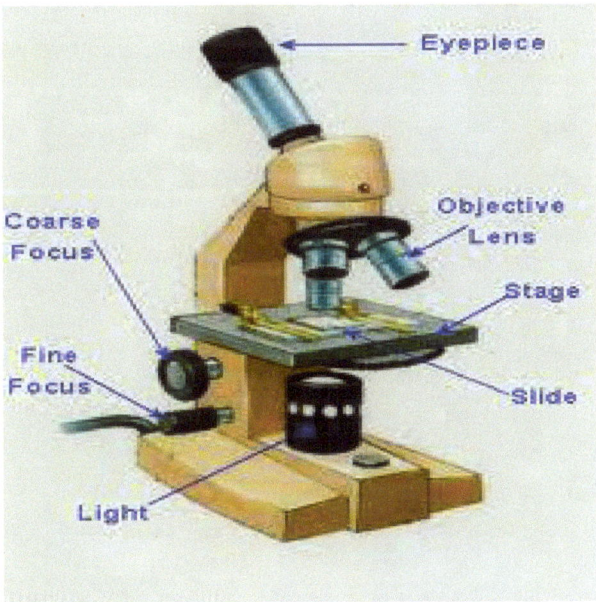

Microscopy reveals cell structure

Three types of microscopes are most commonly used: the compound light microscope, transmission microscope, transmission electron microscope and scanning electronic microscope.

In a compound light microscope, light rays passing through a specimen are brought to a focus by a set of glass lenses, and the resulting image is then viewed by the human eye. In the transmission electronic microscope, electrons passing through a specimen are brought to a focus by a set of magnetic lenses, and the resulting image is projected onto a fluorescent screen or photographic film.

Helminthes

In addition to organism properly in the domain of microbiology which was considered in units 2 to 6, in this unit some microscopic helminthes (worm) will be considered. The helminthes have microscopic stages in their life cycles that can cause disease and arthropods can transmit these stages, as well as other disease-causing microbes.

Complementary texts for reading and translation

Text 1

The animal cell

The animal cell is bounded by a cell membrane, or plasma membrane, which is a triple-layered structure, composed of protein and a lipid (fat-like substance). This membrane is continuous with the cell's internal membrane systems such as the endoplasmic reticulum and the Golgi complex. The great similarity of the membranes of cell organelles of most types thus far investigated has led to the belief that all cell membranes have the same fundamental molecular construction, a concept

Relative sizes of some animal cells and parts of cells. Each major scale division is one-tenth of that above. Visual microscope magnifies about 10 to 2,000 × ; electron microscope about 5,000 to 100,000 × or more.

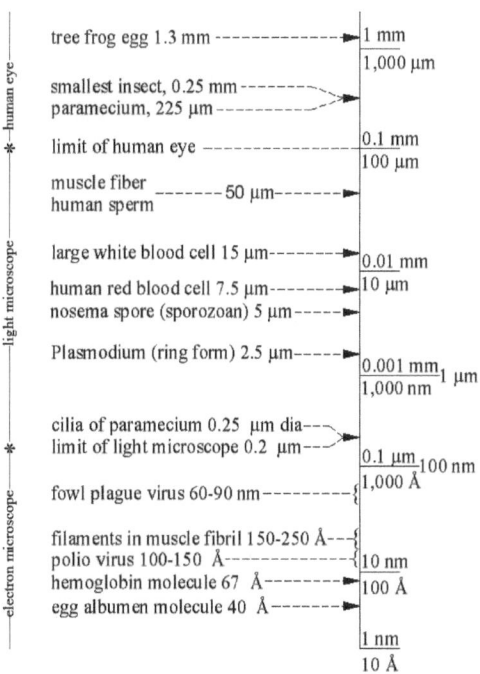

called the unit membrane. The exact molecular arrangement of the protein and lipid molecules in this unit membrane is still unresolved but is believed to be a sandwich of two layers of protein molecules surrounding a layer of lipid molecules. The plasma membrane regulates cell permeability to various kinds of molecules and surrounds the cytoplasm that fills the cell interior. The cytoplasm is translucent and viscous and contains various finer structures and cell organelles ("little organs"). It is surrounded by a distinct nuclear membrane that is continuous with the plasma membrane and is interrupted by nuclear pores that allow the contents to come in contact with the cytoplasm. Within the nucleus is the chromatin (Gr. Chroma, color), seemingly of isolated granules but actually parts of continuously spiraled filaments, the chromonema. During cell division, the chromatin be-comes aggregated as visible rods, the chromosomes, which are capable of self-duplication through successive generations.

Chromosomes are of the greatest biologic importance, because they contain the elements (genes) directing hereditary transmission of characters. The nucleus controls much of cell metabolism; if it is removed, the cell cannot continue normal activities and soon dies. An isolated nucleus cannot form cytoplasm. Each nucleus contains a spherical nucleolus (one or more), involved in nucleoprotein metabolism.

The cytoplasm contains several kinds of structures, cell organelles, some visible under the optical microscope and others shown only by the electron microscope. These organelles and structures are:

1. A spherical centrosome containing one or two dark-staining centrioles. The centrioles have a part in cell division.

2. Golgi complex (bodies or apparatus), often near the centrosome, composed of flattened sacs bounded by membranes continuous with the plasma membrane and thought to be involved in transport of materials in and out of the cell and possibly in certain biochemical reactions requiring membranes for localization of enzymes.

3. Mitochondria, seen as globules or round-ended cylinders or sacs 0.5 to 1 μm in size. They are covered by a membrane about 50 Å thick with an inner membrane folded and projecting into the inner spaces; these inner folds are the site of enzymes directing metabolic oxidations. The mitochondria also contain DNA, which is the genetic material, and a related substance called RNA. The only other Site of DNA in a cell is the nucleus.

4. Endoplasmic reticulum, which is a series of membrane-bounded vesicles of varying shape. The endoplasmic reticulum exists in two types, rough and smooth. Rough endoplasmic reticulum has numerous globular particles 100 to 150 Å in diameter on its inner side. These particles are the ribosome, the sites of protein synthesis. Smooth endoplasmic reticulum lacks ribosome.

5. Microtubules, which appear as long, hollow fibers. They appear to be involved in the preservation of the shape of cells and in the machinery of motion, particularly in mitosis.

6. Lysosomes, which are membrane-bounded bodies containing hydrolytic enzymes.

7. Fat, as droplets or as yolk in egg cells.

8. Vacuoles, or vesicles, small cavities filled with either fluid or granular material.

9. Secretion granules, especially in gland cells, which are transformed to pass out as secretions.

Studies of cells formerly dealt mainly with their physical features as seen in thin stained sections. In recent years, new methods of study and new tools of research have been devised by biochemists to learn the reactions constantly in progress in every living cell. The tiny cell is an amazing unit where many chemical substances undergo a wide variety of interaction and change—synthesis of new materials, use of food and energy to provide for movement, secretion, or other activities, and rendering of waste products into forms not harmful. Any cell is at least as intricate as an entire petroleum refinery that receives the mixture of hydrocarbons in petroleum, refines and modifies some for fuel and lubricants, and synthesizes many new and different organic compounds to serve various purposes in our modern life.

Text 2
CELL DIVISION

Growth in organisms is accomplished chiefly by multiplication of cells. In the unicellular PROTISTA, the animals themselves multiply; in other animals, the number of cells in the individual is increased.

Mitosis cells multiply chiefly by mitosis, a complex process that involves an equal division of the nuclear chromatin in both kind and amount. Cell division by mitosis is common to all animals. Mitosis is active during embryonic development, in growth, in repair of injury, and in replacement of body covering at molting. It is also the process involved in malignant growths (cancer). As seen in living cells, it is a continuous dynamic process, but for study purposes, it is divided into several stages, as follows: (1) prophase, (2) metaphase, (3) anaphase, and (4) telophase. Cells not undergoing mitosis are said to be in the interphase. Duplication of the genetic material occurs in interphase.

PROPHASE The centrosome usually contains two centrioles (if there is only one. this divides): the two move to opposite sides of the nucleus. Around each centriole, fine, short, radiating fibers appear in the cy-toplasm to form an aster; and other longer spindle fibers extend between the separating centrioles.

Meanwhile the chromatin within the nucleus be-comes evident as distinct chromosomes that shorten thicken, and stain deeply. Each chromosome is actually composed of two closely parallel, spiral filaments, the chromatids (daughter chromosomes). In the cells of any one species the several chromosomes are of characteristic size and shape—long or short, thick or thin, and shaped like a rod, a J, or a V. Cereful microscopic preparations show a constriction or dot (centromere) where the two arms of a chromosome join; this is the point of attachment by spindle fibers. Toward the end of the prophase, the nuclear membrane and nucleolus disappear, and the chromosomes become associated with the spindle fibers and move toward the equatorial plane of the cell.

The total number of chromosomes present at the end of the prophase is the diploid number. This is constant and characteristic in any species of animal for all its cells except mature germ cells. In different kinds of animals, the chromosome number ranges from 2 to 250 but usually is less than 50.

METAPHASE   The chromosome lie radially in an equatorial plate across the cell midway between the two asters, each chromosome being connected to the spindle fibers. Other fibers extend continuously between the poles. The two halves of each chromosome become more evident.

ANAPHASE   The halved chromosomes move apart, those of each group toward its respective pole (centriole). In living cells there is an active pulling back and forth of the opposing sets as they separate. Each daughter chromosome consists of an equivalent half of the genetic material formerly in one chromosome.

TELOPHASE   As the groups of daughter chromosomes end their polar movement, they become less conspicuous, a nuclear membrane forms about each group, a nucleolus is produced in each, the centriole divides into two, and the spindle disappears. Finally a cell membrane appears across the former plane of the equatorial plate. When this has ended, the visible part of mitosis is complete. The chromosomes in each daughter cell revert to the net-like pattern of the interphase or metabolic cell.

The equal division of chromatin whereby each daughter cell receives half of that in each parent chromosome is of great significance from the stand-point of heredity, since the genes, or determiners of hereditary characters, are believed to be carried by the chromosomes and to be duplicated with the latter. Such partitioning distributes identical lots of genes to all cells in the body.

Text 3

PHYSICAL PROPERTIES

Matter, weight, and gravity. The substance of the universe, the earth, and living organisms is termed matter. Under different conditions of temperature and pressure, any particular kind of matter may be in one of three physical states—solid, liquid, or gas. Water may be variously solid ice, fluid water, or water vapor. Animal shells and skeletons are mostly of body cells is fluid, and gases are present in lungs or dissolved in body fluids. Almost any animal comprises matter in three states.

The mass or quantity of matter in any object or body is a basic attribute. Certain forces attract any two bodies of matter, the degree of attraction being dependent upon their masses and distance apart. The attraction between the earth and that of any animal or other object on or near its surface is termed gravity, and the value of this force is its weight.

The force of gravity keeps animals against the surface of the earth or any solid object on which they may be. It acts more rapidly in air than in a denser medium, such as water, where resistance to movement is greater. The weight of any given animal would be less on the moon (small mass) but much greater on Jupiter (larger mass). The volume relation of weight of any object in reference to some standard (such as water) is termed its specific gravity. That of a gas is low, whereas that of metals, such as iron or gold is high. Among animals, specific gravity, and particularly the surface-volume relationships, deter-mine their habits and influence the types of environments in which they can live. Bats, birds, and insects are able to fly because of their extensive wing surfaces, and some aquatic invertebrates swim and float readily because they have much surface in relation to weight. The effective specific gravity of any aquatic animal is less than that of a comparable land dweller because the former is buoyed up by the weight of water it displaces. Because of another property or force, inertia, a body at rest tends to remain so, and one in motion tends to continue in motion. Inertia is directly related to mass. A child's wagon requires less force to start into motion (overcoming inertia) than an automobile, but the wagon meets more surface resistance to motion and tends to stop sooner than the heavier vehicle. The same is true of animals. An insect has less inertia than a bear, hence it can start and stop more quickly. In the absence of gravity and friction with the air, water, or ground, a body once set in motion would continue on indefinitely; but on the earth, resistance of the surroundings eventually overcomes the inertia of movement. Any animal, large or small, must exert propulsive power to remain in motion.

Text 4

Kingdoms of organisms

Since the time of Linnaeus until quite recently all organisms were usually classified under one of two kingdoms: Plant or Animal. Plants have photosynthetic pigments and are generally nonmotile, while animals have to obtain food by feeding on plants or other animals and generally move about. Certainly if one concentrates on larger (macroscopic) plants and animals assignment seems easy save, perhaps, for the mushrooms and their relatives (fungi) which are nonphotosynthetic. However, they still retain the plant form and seem-ingly similar growth pattern. Hence, they were assigned to the Plant Kingdom. Major problems with the two-kingdom sequence arose when the microscopic organisms were encountered and closely stud-ied.

Here, for example, we have organisms with all the characteristics of animals and plants, such as Euglena. Furthermore, these microscopic organisms are unicellular, yet the individual cells are more complex than the cells of the bodies of multi-cellular animals. Some microscopic organisms may associate in colonies but do not form tissues as in multicellular animals. The two-kingdom system could not accommodate these organisms without straining the definitions. The result was that botanists claimed motile animals with photosynthetic pigments as plants in their considerations and the zoologists claimed them as animals because they move. In the 1950s and 1960s, biologists began to learn a great deal more about the subcellular structure and biochemistry of cells. Because of this new knowledge, it became increasingly difficult to classify some organisms as either plants or animals. For example, bacteria and blue-green algae organisms have genetic material but no nuclei. Furthermore, these organisms lack the subcellular organelles of the single-celled animals such as mitochondria, vacuoles, and Golgi apparatus. Certain metabolic processes are also different. What should be done with them? They are not animals, and yet they do not have the defined characters of plants.

Another group that does not fit into a two-kingdom system is the fungi. Traditionally grouped with the plants, fungi show some very unplant-like characteristics. For example, plants obtain nourishment through photosynthesis, thereby manufacturing inorganic molecules out of carbon dioxide and water. Fungi do not carry out photosynthesis. Instead, they secrete enzymes into the substrate on which they grow. The enzymes break down complex organic molecules found in animal wastes and dead organisms, and the simpler molecules are then absorbed by the fungus. Another characteristic of fungi is the lack of definite boundaries between cells. As a result, large organic molecules and even nuclei can move from cell to cell. Plants, then, are as different from fungi as they are from one-celled amoebas and bacteria.

As a result of these seemingly fundamental differences, several scientists have proposed dividing the living world into from three to five kingdoms (Table 14-1). The minimum, three, would give the kingdoms PROTOCTISTA , PLANTAE (also called MERA-PHYTA), and ANIMALIA (also called METAZOA). In the three-kingdom system the fungi remain with the plants, while all single-celled organisms are grouped under the PROTOCTISTA. The four-kingdom arrangement would be MONERA, PROTISTA, PLANTAE (MATA-PHYTA),

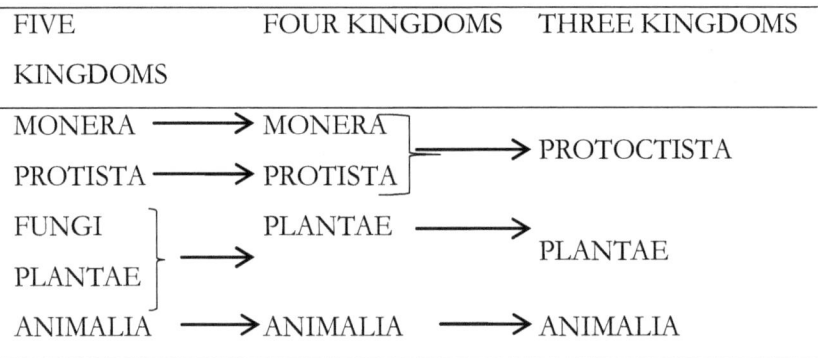

and ANIMALIA (METAZOA). The four-kingdom format divides one-celled organisms into MONERA, which are without nuclei (prokaryotic) and PROTISTA which are with nuclei (eukaryotic). Finally, the five-kingdom system retains the above divisions of the four-kingdom system but makes FUNGI a separate kingdom. Characteristics of the five kingdoms are given in the following table.

Characteristics of five kingdoms of organisms

KINGDOM	TYPE OF CELL	CELL ORGANELLES	NUTRITION TYPE
Monera	Prokaryotic	No membrane around organelles; no plastids; no mitochondria	Absorptive or photo-synthetic
Protista	Eukaryotic	All cell organelles	Absorptive, ingestive photosynthetic
Plantae	Eukaryotic with walls	Present but cells simpler	Photosynthetic mainly
Fungi	Eukaryotic	Lack plastids and photosynthetic pigments	Absorptive
Animalia	Eukaryotic without walls	Lack plastids and photosynthetic pigments	Ingestive

References

Black, J. (2008) (7th edition). *Microbiology*. Asia: John Wiley & sons Pte. Ltd.

Common prefix, suffix and root words. Retrieved 2010 from,

https://www.msu.edu/~defores1/gre/roots/gre_rts_afx2.htm.

Storer, T., Usinger, R. L., Nybakken, J. W., & Stebbins, R. C. (1983). *Elements of zoology*. Tokyo, JIPAn: International Student Edition.

Jawetz, E., & Levinson, W. E. (1993). *Medical microbiology and immunology*. USA: Prentice-Hall International Inc.

Lack, A. J., & Evans, D. E. (2002). *Plant biology*. UK: BIOS Scientific Publishers Limited.

Mader, S. S. (2009) (11th edition). *Human biology*. Boston: McGraw-Hill.

Nussbaum, R. L., McInnes, R. R., & Williard, H. F. (2001) (6th edition). *Thompson & Thompson genetics in medicine*. USA: Saunders Company.

Oxford advance learner's dictionary (2008) (7th edition). Oxford: Oxford University Press.

Plural noun forms. Retrieved 2010 from,

http://grammar.ccc.commnet.edu/grammar/plurals.htm.

Appendix 1

Common root words

Root	Meaning	Examples
acer, acid, acri	bitter, sour, sharp	acerbic, acidity, acrid, acrimony
acu	sharp	acute, acupuncture, accurate
ag, agi, ig, act	do, move, go	agent, agenda, agitate, navigate, ambiguous, action
ali, allo, alter	other	alias, alibi, alien, alloy, alter, alter ego, altruism
alt(us)	high, deep	altimeter, altitude
am, amor	love, liking	amiable, amorous, enamoured
Anni, annu, enni	year	anniversary, annually, centennial
anthrop	man	anthropology, misanthrope, philanthropy
anti(co)	old	antique, antiquated, antiquity
arch	chief, first, rule	archangel, architect, archaic, monarchy, matriarchy, patriarchy, Archeozoic era
aster, astr	star	aster, asterisk, asteroid, astronomy, astronaut
aud, aus	hear, listen	audiology, auditorium, audio, audition, ausculate
aug, auc	increase	augur, augment, auction
auto, aut	self	automobile, autograph, automatic
belli	war	rebellion, belligerent, casus belli, bellicose
bibl	book	Bible, bibliography, bibliomania
bio	life	biology, biometrics, biome, biosphere
brev	short	abbreviate, brief
cad, cas	to fall	cadaver, cadence, cascade
calor	heat	calorie, caloric, calorimeter
cap, cip, cept	take	cIPAble, intercept, forceps, capture, except, reciprocate
capit, capt	head	decapitate, capital, captain, caption
carn	flesh	carnivorous, incarnate, reincarnation, carnal
caus, caut	burn, heat	caustic, cauldron, cauterize
cause, cuse, cus	cause, motive	because, excuse, accusation
ced, ceed, cede,	move, yield, go,	procedure, proceed, cede, concede,

cess	surrender	recede, precede, accede, success
cenetri	center	concentric, centrifugal, centripetal, eccentric
chrom	color	chrome, chromosome, polychrome, chromatic
chron	time	chronology, chronometer, synchronize
cide, cise	cut down, kill	homicide, exorcise, germicide, incision, scissors
cit	call, start	incite, citation, cite
civ	citizen	civic, civil, civilian, civilization
clam, claim	cry out	exclamation, clamor, proclamation, reclamation, acclaim
clud, clus, claus	shut	include, conclude, recluse, claustrophobia, occlusion, occult
cognoac, gnosi	know	recognize, prognosis, cognoscenti, incognito, agnostic
cord, cor, cardi	heart	cordial, concord, discord, courage, encourage
corp	body	corporation, corporal punishment, corpse, corpulent, corpus luteum
cosm	universe, world	cosmos, microcosm, cosmopolitan, cosmonaut
crat, cracy	rule	autocrat, aristocrat, theocracy, technocracy
crea	create	creature, recreation, creation
cred	believe	creed, credo, credence, credit, credulous, incredulous, incredible
cresc, cret, crease, cru	rise, grow	crescendo, concrete, increase, decrease, accrue
crit	separate, choose	critical, criterion, hypocrite
cur, curs	run	current, concurrent, concur, incur, recur, occur, courier, precursor, cursive
cura	care	curator, curative, manicure
cycl, cyclo	wheel, circular	Cyclops, unicycle, bicycle, cyclone, cyclic
deca	ten	decade, decalogue, decathlon, decahedron
dem	people	democracy, demography, epidemic
dent, dont	tooth	dental, denture, orthodontist, periodontal
derm	skin	hypodermic, dermatology, epidermis,

			taxidermy
dict		say, speak	dictation, dictionary, dictate, dictator, Dictaphone, edict, predict, verdict, contradict, benediction
doc, dokein		teach	doctrine, indoctrinate, document, dogma, dogmatic
domin		master	dominate, dominion, predominant, domain
don		give	donate, condone
dorm		sleep	dormant, dormitory
dox		opinion, praise	orthodox, heterodox, paradox, doxology
drome		run, step	syndrome (run together), hippodrome (place where horses run)
duc, duct		lead	induce, seduce (lead aside), produce, reduce
dura		hard, lasting	durable, duration, endure
dynam		power	dynamo, dynamic, dynamite, hydrodynamics
endo		within	endorse, endocardial, endoskeletal, endoskeleton, endosperm
equi		equal	equinox, equilibrium, equipoise
erg		work	energy, erg, allergy, ergometer, ergograph, ergophobia
fac, fact, fic, fect		do, make	factory, fact, manufacture, amplification, confection
fall, fals		deceive	fallacy, falsify, fallacious
fer		bear, carry	ferry, coniferous, fertile, defer, infer, refer, transfer
fid, fide, feder(is)		faith, trust	confidante, fidelity, confident, infidelity, infidel, federal, confederacy, *semper fi*
fila, fili		thread	filigree, filament, filter, filet, filibuster
fin		end, ended, finished	final, finite, finish, confine, fine, refine, define, finale
fix		fix	fix, fixation, fixture, affix, prefix, suffix
flex, flect		bend	flex, reflex, flexible, flexor, inflexibility, reflect, deflect
flu, fluc, fluv		flowing	influence, fluid, flue, flush, fluently, fluctuate
form		form, shape	form, uniform, conform, formulary,

			perform, formal, formula
fort, forc	strong		fort, fortress, fortify, forte, fortitude
fract, frag	break		fracture, infraction, fragile, fraction, refract
gam	marriage		bigamy, monogamy, polygamy
gastr (o)	stomach		gastric, gastronomic, gastritis, gastropod
gen	birth, race, produce		genesis, genetics, eugenics, genealogy, generate, genetic, antigen, pathogen
geo	earth		geometry, geography, geocentric, geology
germ	vital part		germination, germ, germane
gest	carry, bear		congest, gestation
gloss, glot	tongue		glossary, polyglot, epiglottis
glu, glo	lump, bond, glue		glue, agglutinate, conglomerate
grad, gress	step, go		grade, gradual, graduate, progress, graduated, egress
graph, gram	write, written		graph, graphic, autograph, photography, graphite, telegram
grat	pleasing		congratulate, gratuity, grateful, ingrate
grav	heavy, weighty		grave, gravity, aggravate, gravitate
greg	herd		gregarious, congregation, segregate, gregarian
hypn	sleep		hypnosis, hypnotherapy
helio	sun		heliograph, heliotrope, heliocentric
hema, hemo	blood		hemorrhage, hemoglobin, hemophilia, hemostat
here, hes	stick		adhere, cohere, cohesion, inherent, hereditary
hetero	different		heterogeneous, heterosexual, heterodox
homo	same		homogeneous, homonym, homogenize
hum, human	earth, ground, man		humus, exhume, humane
hydr, hydra, hydro	water		dehydrate, hydrant, hydraulic, hydraulics, hydrogen, hydrophobia
ignis	fire		ignite, igneous, ignition
ject	throw		deject, inject, project, eject, interject
join, junct	join		adjoining, enjoin, juncture, conjunction, injunction, conjunction
juven	young		juvenile, rejuvenate

lau, lav, lot, lut	wash	launder, lavatory, lotion, ablution, dilute
leg	law	legal, legislate, legislature, legitimize
levi	light	alleviate, levitate, levity
liber, liver	free	liberty, liberal, liberalize, deliverance
liter	letters	literary, literature, literal, alliteration, obliterate
loc, loco	place	locality, allocate, locomotion
log, logo, ology	word, study, speech	catalog, prologue, dialogue, logogram (symbol representing a word), zoology
loqu, locut	talk, speak	eloquent, loquacious, colloquial, circumlocution
luc, lum, lus, lun	light	translucent, luminary, luster, luna (moon goddess)
macr-, macer	lean	emaciated, meager
magn	great	magnify, magnificent, magnanimous, magnate, magnitude, magnum
man	hand	manual, manage, manufacture, manacle, manicure, manifest, maneuver, emancipate
mand	command	mandatory, remand, mandate
mania	madness	mania, maniac, kleptomania, pyromania
mar, mari, mer	sea, pool	marine, marsh, maritime, mermaid
matri	mother	matrimony, maternal, matriarchate, matron
medi	half, middle, between, halfway	mediate, medieval, Mediterranean, mediocre, medium
mega	great	megaphone, megalopolis, megacycle (a million cycles), megaton
mem	remember	memo, commemoration, memento, memoir, memorable
meter	measure	meter, voltammeter, barometer, thermometer
micro	small	microscope, microfilm, microcard, microwave, micrometer
migra	wander	migrate, emigrant, immigrate
mit, miss	send	emit, remit, submit, admit, commit, permit, transmit, omit, intermittent, mission, missile
mob, mot, mov	move	mobile, motionless, motor
mon	warn, remind	monument, admonition, monitor,

		premonition
mor, mort	mortal, death	mortal, immortal, mortality, mortician, mortuary
morph	form	amorphous, dimorphic, metamorphosis, morphology
multi	many, much	multifold, multilingual, multiped, multiply
nat, nasc	to be from, to spring forth	innate, natal, native, renaissance
neo	new	Neolithic, *nuveau riche*, neologism, neophyte, neonate
neur	nerve	neuritis, neuropathic, neurologist, neural, neurotic
nom	law, order	autonomy, astronomy, gastronomy, economy
nomen, nomin	name	nomenclature, nominate, ignominious
nov	new	novel, renovate, novice, nova, innovate
nox, noc	night	nocturnal, equinox, noctilucent
numer	number	numeral, numeration, enumerate, innumerable
numisma	coin	numismatics
oligo	few, little	Oligocene, oligosaccharide, oligotrophic, oligarchy
omni	all, every	omnipotent, omniscient, omnipresent, omnivorous
onym	name	anonymous, pseudonym, antonym, synonym
oper	work	operate, cooperate, opus
ortho	straight, correct	orthodox, orthodontist, orthopedic, unorthodox
pac	peace	pacifist, pacify, pacific ocean
paleo	old	Paleozoic, Paleolithic, paleomagnetism, paleopsychology
pan	all	Pan-American, pan-African, panacea, pandemonium (place of all the demons),
pater, patr	father	paternity, patriarch, patriot, patron, patronize
path, pathy	feeling, suffering	pathos, sympathy, antipathy, IPAthy, telepathy
ped, pod	foot	pedal, impede (get feet in a trap),

			pedestrian, centipede, tripod, podiatry, antipode
pedo		child	orthopedic, pedagogue, pediatrics
pel, puls		drive, urge	compel, dispel, expel, repel, propel, pulse, impulse, pulsate, compulsory, expulsion, repulsive
pend, pens, pond		hang, weigh	pendant, pendulum, suspend, appendage, pensive
phage		eat	macrophage, bacteriophage
phil		love	philosophy, philanthropy, philharmonic, bibliophile
phlegma		inflammation	phlegm, phlegmatic
phobia, phobos		fear	phobia, claustrophobia, acrophobia, aquaphobia, ergophobia, homophobia
phon		sound	phonograph, phonetic, symphony, homophone, euphonious
photo		light	photograph, photoelectric, photogenic, photosynthesis
plac, plais		please	placid, placebo, placate, complacent
plu, plur, plus		more	plural, pluralist, plus
pneuma, pneumon		breath	pneumatic, pneumonia,
pod (see ped)			
poli		city	metropolis, police, politics, Indianapolis, megalopolis, acropolis
poly		many	polysaccharide, polyandrous, polytheistic
pon, pos, pound		place, put	postpone, component, opponent, proponent, expose, impose, deposit, posture, position, expound, impound
pop		people	population, populous, popular
port		carry	porter, portable, transport, report, export, import, support, transportation
portion		part, share	portion, proportion
pot		power	potential, potentate, impotent
prehendere		seize, grasp	apprehend, comprehend, comprehensive, prehensile
prim, prime		first	primacy, prima donna, primitive, primary, primal, primeval
proto		first	prototype, protocol, protagonist, protozoan, Proterozoic,

		protoindustrial
psych	mind, soul	psyche, psychiatry, psychology, psychosis
punct	point, dot	punctual, punctuation, puncture, acupuncture, punctuation
reg, recti	straighten	regiment, regular, rectify, correct, direct, rectangle
ri, ridi, risi	laughter	deride, ridicule, ridiculous, derision, risible
rog, roga	ask	prerogative, interrogation, derogatory
rupt	break	rupture, interrupt, abrupt, disrupt, ruptible
sacr, sanc, secr	sacred	sacred, sacrosanct, sanction, consecrate, desecrate
salv, salu	safe, healthy	salvation, salvage, salutation
sat, satis	enough	satient (giving pleasure, satisfying), saturate, satisfy
sci, scientia	know	science, conscious, omniscient, cognocienti
scope	see, watch	telescope, microscope, kaleidoscope, periscope, stethoscope
scrib, script	write	scribe, scribble, inscribe, describe, subscribe, prescribe, manuscript
sed, sess, sid	sit	sediment, session, obsession, possess, preside, president, reside, subside
sen	old	senior, senator, senile
senescere	to grow old	senescence, evanescent
sent, sens	feel	sentiment, consent, resent, dissent, sentimental, sense, sensation, sensitive, sensory, dissension
sequ, secu, sue	follow	sequence, consequence, sequel, subsequent, prosecute, consecutive, second, ensue, pursue
serv	save, serve	servant, service, subservient, servitude, preserve, conserve, reservation, deserve, conservation, observe
sign, signi	sign, mark, seal	signal, signature, design, insignia, significant
simil, simul	like, resembling	similar, assimilate, simulate, simulacrum, simultaneous

sist, sta, stit	stand	assist, persist, circumstance, stamina, status, state, static, stable, stationary, substitute
solus	alone	solo, soliloquy, solitaire, solitude
solv, solu	loosen	solvent, solve, absolve, resolve, soluble, solution, resolution, resolute, dissolute
somnus	sleep	insomnia, somnambulist
soph	wise	sophomore (wise fool), philosophy, sophisticated
spec, spect, spic	look	specimen, specific, spectator, spectacle, aspect, speculate, inspect, respect, prospect, retrospective, introspective, expect, conspicuous
sphere	ball, sphere	sphere, stratosphere, hemisphere, spheroid
spir	breath	spirit, conspire, inspire, aspire, expire, perspire, respiration
string, strict	draw tight	stringent, strict, restrict, constrict, boa constrictor
stru, struct	build	construe (build in the mind, interpret), structure, construct, instruct, obstruct, destruction, destroy
sume, sump	take, use, waste	consume, assume (to take, to use), sump pump presumption (to take or use before knowing all the facts
tact, tang, tag, tig, ting	touch	tactile, contact, intact, intangible, tangible, contagious, contiguous
tele	far	telephone, telegraph, telegram, telescope, television, telephoto, telecast, telepathy
tempo	time	tempo, temporary, extemporaneously, contemporary, pro tem, temporal
ten, tin, tain	hold	tenacious, tenant, tenure, untenable, detention, retentive, content, pertinent, continent, obstinate, contain, abstain, pertain, detain
tend, tent, tens	stretch, strain	tendency, extend, intend, contend, pretend, superintend, tender, extent, tension, pretense
terra	earth	terrain, terrarium, territory, terrestrial

test	to bear witness	testament, detest, testimony, attest
the, theo	God, a god	monotheism, polytheism, atheism, theology
therm	heat	thermometer, theorem, thermal, thermos bottle, thermostat, hypothermia
thesis, thet	place, put	antithesis, hypothesis, synthesis, epithet
tom	cut	atom (not cuttable), appendectomy, tonsillectomy, dichotomy, anatomy
tort, tors	twist	torture (twisting to inflict pain) retort, extort, distort, contort, torsion, tortuous, torturous
tox	poison	toxic, intoxicate, antitoxin
tract, tra	draw, pull	tractor, attract, subtract, tractable, abstract, subtrahend
tom	cut	atom (not cutable, smallest particle of matter), appendectomy, tonsillectomy, dichotomy, anatomy
tort, tors	twist	torture (twisting to inflict pain), retort, extort (twist out), distort, contort, torsion
tox	poison	toxic, intoxicate, antitoxin
tract, tra	draw, pull	tractor, attract, traction, subtract, tractable, abstract (to draw away), subtrahend (the number to be drawn away from another).
trib	pay, bestow	tribute, contribute, attribute, retribution, tributary
turbo	disturb	turbulent, disturb, turbid, turmoil
typ	print	type, prototype, typical, typography, typewriter, typology, typify
ultima	last	ultimate, ultimatum
umber, umbraticum	shadow	umbra, penumbra, (take) umbrage, adumbrate
uni	one	unicorn, unify, university, unanimous, universal
vac	empty	vacate, vacuum, evacuate, vacation, vacant, vacuous
vale, vali, valu	strength, worth	equivalent, valiant, validity, evaluate, value, valor
ven, vent	come	convene, intervene, venue, convenient, avenue, circumvent,

			invent, convent, venture, event, advent, prevent
ver, veri		true	very, aver, verdict, verity, verify, verisimilitude
vert, vers		turn	avert, divert, invert, introvert, convertible, reverse, controversy, versatile
vic, vicis		change, substitute	vicarious, vicar, vicissitude
vict, vinc		conquer	victor, evict, convict, convince, invincible
vid, vis		see	video, evident, provide, providence, visible, revise, supervise, vista, visit, vision
viv, vita, vivi		alive, life	revive, survive, vivid, vivacious, vitality, vivisection
voc		call	vocation, avocation, convocation, invocation, evoke, provoke, revoke, advocate, provocative, vocal
vol		will	malevolent, benevolent, volunteer, volition
volcan		fire	volcano, vulcanize, Vulcan
volvo		turn about, roll	revolve, voluble (easily turned about or around *or* talkative), voluminous, convolution
vor		eat greedily	voracious, carnivorous, herbivorous, omnivorous, devour
zo		animal	zoo (short for zoological garden), zoology, zoomorphism (attributing animal form to god), zodiac (circle of animal constellations), protozoan

Glossary

L. = Latin; Gr. = Greek; dim. = diminutive.

Pronunciation is indicated by the division of words into syllables (by hyphens) and by accent marks; on long words with two accented syllables both the primary (") and the secondary (') accents are shown.

Latin and Greek plurals are usually formed as follows: sing. us, to plural i (nucleus, nuclei); a to ae (larva, larvae); um to a (cilium, cilia); on to a (pleuron, pleura); is to es (testis, testes); others are formed by adding s (digit, digits). A few exceptions are genus, genera; species, species; vas, vasa.

Ab-do'men The major body division posterior to the thorax; behind the diaphragm in mammals. (L.)

Ab-duc'tor A muscle that draws a part away from the axis of the body or a limb, or separates two parts. (L. ab, away + duco, lead)

Ab-o'ral Opposite the mouth. (l. ab, from + oris, mouth)

Ab-sorp'tion The selective taking up of fluids or substances in solution by cells or absorbent vessels (L. ab, from + sorbeo, suck in)

Ac-cli'ma-tize To become habituated to an envi-ronment where not native. (L. ad, to + clima, climate)

Ac'i-nus A Small terminal sac in a lung or multicel-lular gland. (L., grape)

Acquired character One that originates during the life of an individual owing to the enviroment or a functional cause.

Ad'ap-ta"tion The fitness of structure, function, or entire organism for a particular environment; the process of becoming so fitted. (L. ad, to + apto, fit)

Ad-duc'tor A muscle that draws a part toward the median axis or that draws parts together. (L. ad, to + duco, lead)

Ad'i-post Pertaining to fat. (L. adeps, fat)

Ad-o'ral Near the mouth (L. ad, toward + oris, mouth)

Ad-sorp'tion Adhesion of an extremely thin layer of gas molecules, dissolved substance, or liquid to a solid surface. Compare Absorption (L. ad, to + sorbeo, suck in)

A-e'ri-al Living or occurring in air. (Gr. aer, air)

Aes'ti-vate Passing the summer in a quiet, torpid condition. (L. aestas, summer)

Af'fer-ent A vessel or structure leading to or toward a given position. (L. ad, to + fero, bear)

Al'bin-ism Lack of pigment when normally present (L. albus, white)

Al'i-men"ta-ry Pertaining to food, digestion, or the digestive tract. (L. alimentum, food)

Al-lan'to-is An embryonic membrane outpocketed from the hindgut and serving for respiration and ex-cretion in embryos of reptiles and birds; becomes part of umbilical cord and unites with the chorion to from the placenta in mammals. (Gr. allas, stem + eidos, form)

Al-le'le The alternative form of a gene, having the same locus in homologous chromosomes; also, the alternative form of a Mendelian character. (Gr. al-lelon, of one another)

Al-le'lo-morph See Allele.

Alternation of generations Metagenesis.

Al-ve'o-lus A small cavity or pit; a tooth socket; a minute terminal air sac in a lung, a terminal unit in an – alveolar gland, one droplet in an emulsion . (L.,a small cavity)

Am'i-no acid Any organic acid containing an amino radical (NH2); amino acids are the "building stones" of proteins.

Am'ni-on The innermost thin double membrane filled with watery amniotic fluid that encloses developing embryo of a reptile ,bird , or mammal .Asimilar single membrane around the insect embryo .(Gr./fetal membrane)

A-moe'boid putting forth pseudopodia like an amoeba or a white blood cell. (Gr.amoibe, change)

Am-phib'i-ous Capable of living either on land or in water, as a frog .(Gr . amphi ,on both sides + bios life)

Am'phi-coe " lous Concave at both ends, as the centrum of some vertebrae . (Gr . amphi , on both sides + koilos , hollow)

Am'phi-mix "is Union of egg and sperm unclei to from a zygote mingling of the germ plasm of two individuals . (Gr. Amphi, on both sides + mixis , a mingling)

Am-pul'la A small, bladder – like enlargement . (L., flask)

A-nab'o-lism Constructive stages in metabolism , in – cluding digestion to assimilation . (Gr. Ana , up + ballo , throw)

A-nal' o-gy Similarity of external features or function, but not of structural plan or origin .(Gr. analogia , ratio)

An-am'nia Vertebrates in which the embryo . is not enclosed by an amnion during development- the cyclostomes fishes and amphibians . (Gr.an,with cyclostomes, fishes , and amphibians .(Gr.an , with out + amnion)

A-nas'to-mo" sis A union or joining as of two or more arteries , veins , or other vessels . (Gr. ana , again + stoma , mouth)

An-ten'na A sensory appendage, especially on arthropods, not concerned with light perception or sight .(Gr. ana , up + teino , stretch)

An-ter'i-or The forward-moving or head end of an animal,or toward that end . Opposite of posterior . (L. onte , before)

An'ti-mere One of the several similar or eguivalent parts into which a radially symmetrical animal may be divided . (Gr. ana , against + meros, part)

A'nus The posterior opening of the digestive tract (L .)

A-or'ta A large artery , cspecially one connected to the heart . (Gr.aorte,artery)

A-or'tic arch A large artery arising from the heart in vertebrates ; one of paired arteries connecting the ventral aorta and dorsal aorta in the region of the pharynx or gills.

Ap'i-cal At the apex or top, as of a conical structure. (L.)

Ap-pend'age A movable projectiong part on a meta zoan body having an active function .(L. ad, to + pendeo, hang)

A-guat'ic Pertaining to or living in water .(L. apua, water)

Ar-bor'e-al Pertaining to or living in trees , as tree inhabiting animals . (L.arbor, tree)

Ar-chen'ter-on The primitive digestive cavity of a metazoan embryo , formed by gastrulation . (Gr.archi,first + enteron , intestine)

Ar'ter-y A tubular vessel conveying blood away from the heart. (Gr.arteria, artery)

Ar-tlc'u-late to attach by a joint . (L.articulus, joint)

A-sex'u-al Not related to sex ; not involving gametes or fusion of their nuclei.

As-sim'i-la"tion Incorporation of digested nutria ment, after absorption , into living protoplasm. (L.ad, to + similes , like)

A'sym-met" ri-cal Without symmetry .

A'tri-um An outer cavity or chamber , the receiving chamber of the heart. (L . court)

Au"di-to'ry Pertaining to the organ or sense of hearing . (L.audio,hear)

Au'ri-cle The external pinna of the ear in mammals .(L.)

Au'to-some Any ordinary chromosome as contrasted with a sex chromosome. (Gr.autos, self + soma, body)

Au-tot' o-my The automatic "voluntary" breaking off of a part by an animal. (Gr. autos,self + temno , to cut)

Au'to-tro" phic nu – tri'tion That process by which an organism manufactures its own food from in organic compounds, as in a plsnt . (Gr. autos , self + trepho , to feed)

Ax ' i – al skel 'e-ton That part of the vertebrate skele ton in the sxis of the body—skull , vertebrae , ribs, and sternum.

Ax'is A line of reference or one about which parts are arranged symmetrically. (L.)

Ax'on . Ax' one The process of a nerve cell that con ducts impulses away from the cell body of which it is a part. (Gr.,axon , axle)

Bi-lat ' er-al sym'me-try Symmetry of a kind so that a body or part can divided by one median plane into equivalent right and left halves , each a mirror image of the other .

Bi-ra'mous Consisting of or possessing two branches , as a crustacean appendage (L.bis,twice + ramus, branch)

Blad'der A thin-walled sac , or bag , that contains fluid or gas.

Blas'to-disc The germinal area on a large yolked egg that gives to the embryo . (Gr. blastos, germ + diskos , platter)

Blas'to-mere One of the early cells formed by the division of an ovum . (Gr.blastos,germ+meros,part)

Blas'TO-PORE The mouth – like opening of a gastrula .(Gr.blastos, germ + poros , passage)

Blas'tu-la Early stage of an embryo, usually a hollow sphere of cells .(Gr.dim.of blastos, germ)

Blood The fluid that circulates in the vascular or circulatory system of many animals.

Bod'y cav'i-ty the cavity between the body wall and internal organs of an animal. see coelom, pseudocole.

Bra'chi-al Referring to the fore limb or pectoral appendage. (L.brachium,forearm)

Bran'chi-al Referring to gills. (Gr.branchia,gills)

Bron'chus One of the larger divisions of the trachea conveying air into the lungs. (G . bronchos , windpipe)

Buc'cal Pertaining to the mouth or cheek. (L . bucca , cheek)

Bud Part of an animal that grows out to produce a new individual.

Bur'sa A pouch,or sac , as the bursa of a joint . (Gr. hide or skin)

Cal-car'e-ous Composed of or containing calcium carbonate ($caco_3$).(L . calx or calc, lime)

Cal' o-rie Unit of heat. Small calorie , amount needed to raise 1 g of water 1c. large calorie, amount necessary to raise 1 kilogran of water 1c(at 15c).

Cap'il-la-ry A minute tubular vessel with walls composed of a single layer of thin cells, through which diffusion may occur ; commonly in a connecting network between arteries and veins . (L. capillus , hair)

Car' a-pace The hard shell of turtles and crustaceans . (Spanish , carapacho)

Car'di-ac Pertaining to or near the heart. (G.r. kardia , heart)

Car-niv' o-rous Eating or living on flesh of other animals .(L . caro , flesh + voro, eat)

Cas-tra'tion Removal of the gonads, or sex glands, especially of the male. (L. castro ,to castrate)

Ca-tab'o-lism Destructive metabolism, the breaking down of more complex substances in protoplasm. (Gr. kata , down + ballo, throw)

Cau'dal pertaining to the tail, or posterior past of the body. (L. cauda , tail)

Ce'cum; pl . ce'ca A pouch or sac-like extension on the digestive tract, closed at the outer end. (L. caecus, blind)

Cell A small mass of living matter usually containing a nucleus or nuclear material, the fundamental unit of structure and function in plants and animals . (L. cella, small room)

Cel'lu-lar Pertaining to or consisting of cells. (L.dim . of cella, small room)

Cel'lu-lose The carbohydrate forming the wall of plant cells, also in the mantle of tunicates.

Cen'trum   The spool – like body of a vertebra, which bear various processes. (L. center)

Ce-phal'ic   Pertaining to or toward the head. (Gr. kephale, head)

Ceph'a-lo-tho " rax   A body division with the head and thorax combined. (Gr. kephale, head+ thorax, chest)

Cer'e-bel"um   The anterior development from the hindbrain. (L. dim. Of cerebrum, brain)

Cer'e-bral   Of or pertaining to the brain as a whole or the anterior dorsal (cerebral) hemispheres; also to the anterior brain – like nerve ganglia of various invertebrates . (L. cerebrum, brain)

Cer'e-brum   The dorsal anterior part of the vertebrate forebrain , consisting of two "hemispherical" masses. (L. brain)

Cer'vi-cal   Pertaining to a neack . (L. cervix , neck)

Chae' ta   See seta.

Char' ac- ter, Characteristic   A distinguishing feature , trait, or property of an organism. (Gr.)

Che- lic' e- ra   One of the most anterior pair of appendages on arachnoids as spiders, scorpions , and the horseshoe crab. (Gr. chele , claw + keras; horn)

Che'li-ped   The first thoracic appendage (pincer) a crayfish and related crustaceans . (Gr. chele, claw + L. pes , foot)

Chi'tin   The structural carbohydrate secreted in the exoskeleton on arthropods and some other animals. (Gr. chiton, tunic)

Chlo'ro-phyll   The green pigment of plants and certain animals, involved in photosynthesis . (Gr. chloros, green + phyllon, left)

Cho'a-na   A funnel, especially the opening between the nasal passages and pharynx (or mouth).(Gr.)

Chon'dro - cra" ni- um   The cartilaginous skull of cyclostomes and elasmobranches; also that part of the embryonic skull in higher verebrates first formed as cartilage. (Gr. chondros , cartilage + kranion , skull)

Chor - da ' ta   The phylum of animals wit a notochord , persistent or transient , includes the vertebrates amphioxus , and tunicates; the chordates. (L.chorda, cord or string)

Cho'ri-on   The outer double membrane surrounding the embryo of a reptile , bird, or mammal; in mammals it unites with the allantois to form the placenta ; the outer membrane of an insect egg. (Gr. membrane)

Chro'ma-tin   The easily and deeply staining substance in a cell nucleus, conspicuous in the nuclear network and in the chromosomes at mitosis . (Gr. chroma , color)

Chro"ma-to-phore'   A pigment cell containing granules or coloring material and responsible for color markings on many animals (Gr. chroma, color + phero , bear)

Chro'mo-mere   Dark-staining localized thickenings in a chromosome. (Gr. chroma , color + meros, part)

Chro'mo-somes   characteristic deeply staining bodies , formed of chromatin in the nucleus af a cell during mitosis, that bear the genes or determiners of heredity . (Gr. chroma, color + soma, bady)

Cil'i-um;   pl.cil'i-a A microscopic hair-like process attached to a free cell surface ; usually numerous, often arranged in rows, and capable of vibration. (L.eyelid)

Cir'rus;   pl. cir'ir A small , slender, and usually flexible structure or appendage . (L. tuft of hair)

Cleav' age   The early stages in the division of an egg cell into many cell.

Clo-a'ca   The terminal portion of the digestive tract in many insects; the common passage from the digestive , excretory, and reproductive organs in various vertebrates. (L. sewer)

Co-coon'   A protective case or covering about a mass of eggs, a larva or pupa, or even an adult animal.

Coe'lom   The body cavity or space between the body wall and internal organs in many metazoan animals, lined with peritoneum (mesoderm). (Gr. koilos, hollow)

Coe-lom'o-ducts   Ducts, derived from mesoderm, that convey gametes or excretory products (or both) from the coelom to the exterior.

Col'o-ny   A group of organisms of the same species living together; colonial . Opposite of solitary. (L.colonus, farmer)

Com-men'sal-ism   The association of two or more individuals of different species in which one kind or more is benefited and the others are not harmed . (L.cum, together + mensa, table)

Com-mu'in-ty   A group of organisms of one or more species living together and related by environmental requirements . (L. communis , common)

Com-pressed'   Reduced in breadth, flattened laterally. (L. cum , together + premo , press)

Con-ver'gent Approaching each other or tending toward a common point. (L. con, together + uergo, incline)

Con' vo – lut" ed Coiled or twisted. (L.conuolo, roll together)

Cop' u-la"tion Sexual union. (L.copulo, join together)

Co'r-ium The dermal portion of the skin beneath the epidermis. The dermis. (L.skin, hide)

Cor'ne-a The outer transparent coat of an eye. (L.comeus, horny)

Cor-ni-fied' Hardened or horn – like, such as a callus on skin, a nail or claw, a bird'd bill, etc. (L.comeus, horny)

Cor'pus-cle A small or minute structure or a cell free or attached, as a blood corpuscle or bone corpuscle. (L.dim. of corpus, body)

Cor'tex The outer or covering layer of a structure. (L.rind, bark)

Cra'in-al Of or pertaining to the skull or brain, as a cranial nerve. (Gr. kranion, skull)

Cra'ni-um The skull, specifically the brain case. (G. kranion, skull)

Crop A thin – walled and expanded portion of the digestive tract, primarily for food storage.

Cross-fer'til-i-za"tion Union of an egg cell from one individual with a sperm cell from another. Opposite of self-fertilization.

Cu-ta'ne-ous Pertaining to the skin. (L. cutis, skin)

Cu'ti-cle A thin, noncellular external covering on an organism. (L. dim. of cutis, skin)

Cyst A resistant protective covering formed about a protozoan or other small organism during unfavorable conditions or reproduction, a small sac or capsule. (Gr. kystis, bladder)

Cy'to-plasm That part of a cell outside the nucleus and within the cell membrane. (G. kytos, hollow + plasma, form)

Cy'to-some See cytoplasm.

Def' e-cate To discharge food residues (feces) through the anus. (L.de, from + faex, dregs)

Den'drite The process on a nerve cell that conducts impulses to the cell body, often branched. (Gr. den-dron, tree)

De-pressed' Flattened vertically, from above (L.de, down + premo, press)

Der'mal Pertaining to the skin, especially the inner connective tissue layers of vertebrate skin.(Gr.derma, skin)

Der'mis The deeper or "true" portion of the skin beneath the epidermis in a vertebrate, derived from mesoderm. (Gr. derma,skin)

Di'a-phragm A dividing membrane, as the diaphragm of the ear; the muscular partition between the thoracic and abdominal cavities in mammals . (Gr.dia, through + phragma, fence)

Di-ges'tion The process of preparing food for absorption and assimilation.(L.)

Dig'it A finger or toe; one of the terminal divisions of a limb in tetrapods. (L.digtus, finger)

Dig'i-ti-grade Walking on the toes. (L.digitus, finger + gradior, to step)

Di-mor'phism Existing under two distinct forms. (Gr. di, two + morphe, form)

Di-oe'cious With the male and female organs in separate individuals .(Gr.di, two + oikos, house)

Dip'lo-blas"tic Derived from two embryonic germ layers, ectoderm and endoderm. (Gr. diplous, double + blastos, germ)

Dip'loid The dual or somatic number of chromosomes (2n) the normal number in all but the matured germ cells of any particular organism. (Gr. diplous, double + eidos, form)

Dis'tal Away from the point of attachment or place of reference. (L. disto, stand apart)

Di-ur'nal Pertaining to, or active during, the daytime. (L.dies, day)

Di-ver'gent Going farther apart; separating from a common source. (L. di, doubly + uergo, incline)

Dom'i-nant char'ac-ter Acharacter from one parent that manifests itself in offspring to the exclusion of a contrasted (recessive) character from the other parent. (L.dominor, rule)

Dor'sal Toward or pertaining io the back, or upper surface. (L. dorsum, back)

Duch A tube by which a liguid or other product of metabolism is conveyed, as of a secretion from a gland, usually opening on a surface or in a larger compartment. (L.duco, lead)

Duct'less gland A gland that elaborates and secretes a hormone, or "internal secretion," directly into the bloodrtream; an endocrine gland.

Duc'tus de'fer-ens The sperm duct from the efferent ductules to the cloaca or ejaculatory duct. (L.deferens, carry out)

E-col'o-gy The relations of an organism to its environment. (Gr. oikos, home + logy)

Ec'to-derm The outer germ layer or cell layer of an early embry . (Gr. ektos, outside + derma, skin)

Ec'to - par" a - site One that lives on the exterior of its host.(Gr. ektos, outside + parasite)

Ec' to-ther " mal Deriving body heat from the environment, characteristic of all animals but birds and mammals.(Gr. ekto, to or on the outside + thermal)

Ef – fec' tor A structure that transforms motor impulses into motor action. (L. efficio, effect, bring to pass)

Ef'fer-ent A structure leading away from a given point of reference, as an efferent artery. (L. exout + fero, carry)

Ef'fer- ent duc' tules Short ducts carrying sperm from testis to ductus deferens.(L.ex, out + fero, carry + duco, lead)

E- gest' To discharge unabsorbed food or residues from the digestive tract . (L.e, out + gero, bear)

Egg A germ cell produced by a functionally female organism, an ovum .

Em' bry-o A newly forming younf animal in the stages of development before hatching or birth. (Gr.)

Em' bry- og" e-ny The process of development of the embryo. (Gr.embryon + genesis , generation)

Em' bry –on" ic mem ' branes Cellular membranes formed as part of an embryo during its development and necessary for its metabolis; the amnion chorion , and allantois of reptiles, brds, and mammals, and of insects.

En-am'el The dense whitish covering on teeth of vertebrates, the hardest substance produced by animal bodies.

En' do-crine A ductless gland; also, its internal secretion or hormone, which is diffused into and carried by the bloodstream. (Gr.endon, within + krino, separate)

En' do -derm (or **entoderm**) The layer or group of cells lining the primitive gut, or archenteron, in an early embryo, beginning in the gastrula stage.(Gr. endon, within + derma, skin)

En-dog' e-nous Growing or originating from within. (Gr.endon, within + gigno, be born)

En' do-par "a-site One that lives within its host. (Gr. endon, within + parasite)

En' do-skel" e-ton An internal supporting framework or structure.(Gr. endon , within + skeleton)

En' do-style The ventral ciliated groove in the pharynx of tunicates, amphioxus, and larval lampreys, used in food getting; homologous with the thyroid gland of vertebrates. (Gr. endon , within + stylos, column)

En'd0-the"li-um Layer of simple sguamous cells lining the inner surface of circulatory organs and other closed cavities. (Gr.endon, within + thele, nipple)

En' do-ther"mal Generating body heat from within; birds and mammals. (Gr.endon, within + termal)

En 'ter-on The digestive cavity, especially that part lined by endoderm. (Gr. intestine)

En-vir 'on -ment The total of conditions surrounding an entire organism.

En' zyme A substance produced by living cells that in minute amount causes specific chemical transformation such as hydrolysis, oxidation, or reduction but that is not used up in the process; a ferment or catalyst. (Gr. en, in + zyme, leaven)

Ep'i- der" mis A layer of cells (sometimes stratified) covering an external surface; the ectodermal portion of the skin of most animals; secretes cuticle on some animals. (Gr. epi, upon + derma, skin)

Ep'i-did" y-mis The efferent tubules of the testes. (Gr. epi, upon + didymos, testicle)

E-piph' y-sis The and and or other external part of a bone that ossifies separately; also , the pineal body , a dorsal outgrowth on the diencephalons of the vertebrate brain. (Gr. epi, upon + phyto, to grow)

Ep'i-the" li-um A layer (or layers) of cell covering a surface or lining a cavity. (Gr. epi, upon + thele, nipple) red blood cell or corpuscle; characteristic of vertebrates. (Gr. erythros , red + kytos, hollow vessel)

E-ryth 'ro -cyte A red blood cell or corpuscle; characteristic of vertebrates . (Gr. erythros, red + kytos, hollow vessel)

E-soph 'a - gus That part of the digestive tract between the pharynx and stomach .(Gr.)

Eu-sta' chi -an tube The passage between the pharynx and middle ear in land vertebrates. (Eustachio, on ltalian anatomist)

E-vag 'i-na " tion An outpocketing from a hollow structure. (L. e, out from + uagina , a sheath)

E'vo -lu "tion The process by which living organisms have come to be what they are, structurally and functionally, complex forms being derived from simpler form; hence, descent with modification. (L. evolvo, unroll, unfold)

Ex-cre 'tion Waste material resulting from metabolism and discharged from the body as useless; also the process of its elimination. Compare secretion, and feces.(L. excemo, separate, secrete)

Ex'o –skel "e-ton An external supporting structure or covering .(Gr. exo, outside + skeleton)

F1,F2,etc. Abbreviations for 1st filial, 2d filial, etc. indicating the successive generations following crossbreeding.

Fac' tor An agent or cause; in genetics; a specific germinal cause of a hereditary character ; same as gene .

Fas' ci-a A sheet of connective tissue covering an organ or attaching a muscle. (l;a band)

Fau' na All the animal life in a given region or period of time.

Fe' ces Excrement ; unabsorbed or indigestible food residues discharged from the digestive tract as waste. (l;dregs)

Fer' til- i- za "tion Union of two gametes (egg and sperm) to form a zygote and initiate the development of and embryo. (L.fertil, fruitful, from. Fero, to bear)

Fe' tus The later stages of and embryo while within the egg or uterus.(L. offspring)

Fi' ber A delicate, thread-like part in a tissue. (L.)

Fi' bril A small fiber. (L.)

Fin An extension of the body on an aquatic animal, used in locomotion or steering.

Fis' sion Asexual reproduction by division into 2 or more parts, usually equivalent.(L. findo, split)

Fla-gel' lum A long lash or thread-like extension capavle of vivration .on flagellate protozoans and on collar cells of sponges.(L. little whip)

Flame cell A type of hollow terminal excretory cell in certain invertebrates that contains a beating (flame-like) group of cilia.

Fol ' li-cle A minute cellular sac or covering .(L.dim.of follies, bag)

Food vac ' u-ole An intracellular digestive organelle.

Fo – ra'men An opening or perforation through a bone, membrane, or partition. (L. foro, bore)

Fos' sil Any relic of an organism buried in the earth or rocks by natural causes in past geologic time.(L. fodio , dig)

Free- liv' ing Not attached or parasitic; capable of in dependent movement and existence. Compare Sessile.

Fron' tal Of or pertaining to the front, or forehead; also , a plane or section parallel to the main body axis and at right angles to the sagittal plane (L. frons, the brow)

Func' tion The activity or action of any part of and organism .(L. functio, to perform)

Gam' ete A mature reproductive or germ cell, either male (sperm) or female (ovum).(Gr.ganos,marriage)

Gam' e- to- gen" e- sis The process of formation of mature germ cells, or gametes; maturation.

Gan' gli- on; pl. gan' gli- a A group or concentration of nerve call bodies , set apart, and acting as a center of nervous influence.

Gas' tro- coel The primitive digestive cavity of a metazoan embryo, formed by gastrulation . (Gr. gaster, stomach + koilos, hollow)

Gas' tro- der" mis Lining of the digestive cavity in coelenterates. (Gr. gaster, stomach + derma, skin)

Gas' tro- vas" cu- lar Serving for both digestion and circulation.

Gas' tru- la Early stage in embryonic development; an invaginated blastula.(Gr. gaster, belly)

Genes The units of inheritance, which are transmitted from one generation to another in the gametes and control the development of characters in the new individual; the factors, or hereditary determiners.

Gen' i – tal Referring to reproductive organs or the process of generation. (L. gigno, beget)

Gen' o type The internal genetic or hereditary constitution of an organism without regard to its external appearance. Compare phenotype. (Gr. genos, race + typos , impression, form)

Germ cell A reproductive cell in a multicellular organism.

Germ lay' er One of the (two or three) fundamental cell layers (ectoderm endoderm, mesoderm) in an early embryo of a multicellular animal from which tissues and organs of the adult are formed.

Germ plasm The material basis of inheritance, the gametes and the cells and tissues from which they from, considered as a unit.

Ger' min – al va' ir- a "tions Those due to some modification in the germ cells.

Gill An organ for aquatic respiration.

Giz' zard A heavily muscled portion of the digestive tract.

Gland An organ of secretion or excretion .(L. glans, acorn)

Glo- mer' u- lus A small, rounded clump of vessels; the knot of capillaries in a renal corpuscle. (L. dim. Of glomus, ball)

Glot' tis The opening from the pharynx into the trachea. (Gr. glotta, tongue)

Gly' co- gen A carbohydrate (polysaccharide) stored in the muscles and liver; "animal starch , "(Gr. plykys, sweet + gen, come into being)

Gon' ad A reproductive organ (ovary, testis, or hermaphrodite) in which gametes (ova or sperm) are produced. (Gr. gonos , generation, seed)

Gre- ga' ri-ous Habitually living in groups, flocks, etc, of numerous individuals. (L. grex, herd)

Gy- nan' dro-morph An individual in a dioecious species having one part of the body fermale and another part male in constitution. (Gr. gyne, woman + aner, man)

Hab' i-tat The natural or usual dwelling place of an individual or groups of organisms . (L. habitus, condition)

Hair A slender filamentous growth on the skin of mammals and on the exposed surfaces of some arthropods.

Hap' loid The sigle or halved number of chromo-somes (n) as found in matured germ cells . (Gr. hap lous, single + eidos, from)

Hem' al Pertaining to the blood or the blood –vascular system. (Gr. aima, blood)

Hem' o- coel Portion of a body cavity reduced in size and functioning as part of a blood –vascular system. (Gr. aima, blood + koilos, cavity)

Hem' o- glo" bin The coloring matter of red corpuscles in vertebrate blood and blood plasma in some in vertebrates; a protein containing iron that combines with and transports oxygen to the tissues. (Gr. aima, blood + globe)

He- pat' ic Ppertaining to the liver. (Gr.hepar, liver)

He- pat' ic por' tal sys' tem A system of veins leading from the digestive tract to capillaries (sinusoids)in the liver of a vertebrate.

Her- biv' o- rous Feeding only or chiefy on herbs,grasses , or other vegetable matter. (L.herba,grass +uoro , devour)

He – red" i- tar' y Passing by inheritance from one generation to another.

He- red' i- ty The transmission of characters, physical and others, from parent to offspring; the tendency of offspring to resemble their parents. (L.heres, heir)

Her- maph' ro- dite An animal with both male and female sex organs. (Gr.Hermes +Aphrodite)

Het' er- o- zy" gote An individual produced by union of two germ cells that contain unlike genes for a given character, either both genes of an allelomor – phic pair or two different genes of an allelomorphic series. Compare Homozygote. (Gr. heteros, an- other +zygon, yoke)

Hi' ber- nate Passing the winter in an inactive or torpid condition. (L. hibemo, pass the winter)

Hol' o- blas" tic Cleravage in which an entire egg cell divides. (Gr. holos, whole + blastos, germ)

Hol' o- phyt" ic nu- tri' tion Nutrition involving pho- tosynthesis of simple inorganic chemical substances, as in green plants and some flagellate protozoans. (Gr. holos, whole + phyton, plant)

Hol' o-zo" ic nu-tri' tion Nutrition requiring com – plex organic foodstuffs, and characteristic of most animals. (Gr.holos, whole + zoon, animal)

Ho' moin-ther" mal Having constant internal tem- perature, often maintained above that of the environ –ment ; characteristic of birds and mammals. (Gr. ho- moios, like + themal)

Ho- mol' o-gous Of like source in structure and embryonic development from primitive origin .(Gr. homos, same + lego, speak)

Ho- mol' o-gous chro' mo- somes A pair of chromo – somes having relatively similar structure and value ,one from each parent. (Gr. homologos, agreeing)

Ho- mol' o-gy Fundamental similarity; strucyural likeness of an organ or part in one of animal with the comparable unit in another resulting from descent from a common ancestry .Compare Analogy. (Gr.)

Ho' mo- zy" gote An individual produced by union of two germ cells that contain like genes for a given character. Compare Heterozygote. (Gr. homos, like + zygon, yoke)

Hor' mone A chemical regulator or coordinator secreted by cell or ductless glands and carried in the bloodstream. see Endocrine. (Gr. hormao, urge on, spur)

Host An organism that harbors another as a parasite. (L. hospes, entertainer)

Hy' a-line Glassy or semitransparent . (Gr. hyalos, glass)

Hy' brid The offspring of two parents that differ in one or more heritable characters; a heterozygote. (L. hybrida, mongrel)

Hy- per' tro- phy Abnormal increase or overgrowth in the size of a part or organ . (Gr. hyper, over + trepho, nourish)

In- breed' To mate related animals or plants.

In- gest' To take food into a place of digestion. (L. in + gero, bear)

In- her' i- tance The sum of all characters that are transmitted by the germ cells from generation to generation. (L. in, in + heres, heir)

In' stinct An inherited type of action, invoked by a certain stimulus and often of complex nature, combining associated reflex acts and leading to a particular end.

In- teg' u-ment An outer covering especially the skin of a vertebrate and its derivatives. (L. intego, to cover)

In' ter- cel" lu- lar Between or among cells.

In- tes' tine That part of the digestive or alimentary canal between the stomach and anus (or cloaca). (L. intus, inside)

In' tra- cel" lu-lar Within a cell or cells.

In- vag' i- na" tion An inpocketing or folding in, as of the vegetal pole of a blastula to form the gastrula. (L. in + vagina, a sheath)

In- ver' te- brate Any animal without a dorsal column of vertebrae: protozoans to amphioxus, inclusive.

Ir' ri- ta- bil" i- ty The capacity of responding to stimuli. (L. irrito, excite)

Joint A place of union between two separate bones or other hardened structures: also to segments of arthropod appendages: 6-iointed, ect. (L. junctus, from jungo, to join)

Kar' yo- type Appearance (size, shape, and number)of the set of chromosomes of a somatic cell.

La' bi- al Pertaining to the lips.

La- mel' la A thin, sheet-like layer. (L. lamina, plate)

Lar' va The early and usually active feeding stage of an animal, after the embryo, and unlike the adult. (L. mask)

Leu' ko- cyte A white blood cell or corpuscle. (Gr. leukos, white + kytos, hollow vessel)

Lim'y Calcareous; containing calcium salts, especially $CaCo_3$.

Lin' gual Pertaining to the tongue. (L. tongue)

Link' age Inheritance of characters in groups, probably because their genes lie in the same chromosome.

Lip' oid Of fatty nature. (Gr. lipos, fat)

Lo' pho-phore A ridge about the mouth region bearing hollow tentacles in some invertebrates. (Gr. lophos, crest + phero, bear)

Lu'men The cavity in a gland, duct, vessel, or organ.

Lu' mi-nes" cence. Emission of light as a result of chemical reactions within cells . (L. lumen, light)

Lung An organ for aerial respiration.

Lymph Colorless blood fluit (without red blood cells) found among tissues and in lymph capillaries or vessels. (L. lympha, water)

Lym- phat' ic sys' tem A system of delicate vessels in vertebrates that lead from spaces between tissues to large veins entering the heart; part of the circulatory system.

Lymph' o-cyte A white blood cell with one large undivided and nongranular nucleus; present in blood and lymph vessels. (L. lympha, water + Gr. kytos, hollow vessel)

Man' di- ble The lower jaw of a vertebrate; either jaw of in arthropod. (L. mandibula, jaw)

Ma- rine' Pertaining to or inhabiting the sea, ocean, or other salt waters. (L. mare, sea)

Ma' trix Intercellular substance as in connective tissues, cartilage, etc .

Mat'u-ra" tion Final stages in preparation of sex cells for mating with segregation of homologous chromosomes so that each cell or gamete contains half the usual (diploid) number .

Me- du' sa One of two types of individuals or morphs found in cnidarians. Usually free-swimming and reproducing sexually. (Gr. myth)

Mei – o' sis Special type of cell division which reduces the chromosome number to half . Occurs in sex cells.

Mem' brane A thin ,soft sheet of cells or of material secreted by cells.

Mer'0 – blas" tis Cleavage of an egg in which only part of the protoplasm divides ,leaving the yolk undi-vided ;characteristic of eggs with much yolk. (Gr.meros,part +blastos,gem)

Mes' en- chyme Part of the mesoderm in a vertebrate embryo that produces connective and circulatory tissues. (Gmesos , middle +chyma ,fluid)

Mes' en- ter- y The sheet of tissue that suspends organs in the body cavity and is continuous with the peritoneum lininig that cavity . (Gr.mesos, mid- dle + enteron , intestine)

Mes' o-derm The embryonic cells or cell layers between ectoderm and endoderm. (Gr.mesos, middle +dema, skin)

Mes' o-gle" a The gelatinous filling between the outer and inner cell layers of a two-layered animal such as a jellyfish . (Gr. mesos, middle + glois, glutinus)

Me- tab' o-lism The sum of the constructive and destructive processes (anabolism and catablism) , mainaly chemical , that occur in living organisms. (Gr. metabolos, changeable)

Met' a-gen" e- sis Alternation of sexual and asexual reproduction in the life cycle of certain animals; alternation of generations . (Gr. meta, after + genesis, origin)

Met' a-mere Any one of a series of homologous parts in the body as with annelids, arthropods, or chordates; a somite . (Gr. meta, after + meros, part)

Me- tem' er-ism Segmental repetition of homologous parts (metameres).

Met' a-mor" pho-sis Marked change in form from one stage of development to another, as of a larva to an adult. (Gr.)

Met' a-ne-phrid" i-um A tubular excretory organ the open inner and draining from the coelom and the outer discharging to the exterior, as in the earthworm. (Gr. meta, after + nephros, kidney)

Met' a-zo" a Multicellular animals with cells usually arranged into tissues; includes all animals above the sponges. (Gr. meta, after + zoon, animal)

Mi' cro-meter (micron ; pl. mi'cra).The unit of microscopic measurement, 1/1,000 of a millimeter; represented by um (Greek letter mu).(Gr. micros, small)

Mi-to' sis Cell division that is characterized by the appearance of a fibrous and a definite number of chromosomes, which split longitudinally to form two equal sets of daughter chromosomes; the latter diverge to opposite poles of the spindle to become parts of two new nuclei . (Gr. mitos, thread)

Mo' lar The posterior permanent teeth of a mammal. (L. molo, grind)

Molt To cast off an outer covering such as cuticle, scales, feathers, or hair.

Mo-noe' cious Having both male and female gonads in the same individual; hermaphroditic. (Gr. monos, single + oikos, house)

Mon' o-hy " brid The offspring of parents differing in one character . (Gr. monos, single + L. hybrida, mongrel)

Mon' o-phy-let" ic From a single known evolutionary derivation. (Gr. monos, single + phule, tribe)

Mu' cous Secreting mucus or similar sticky, slimy substance as by a mucous cell gland or membrane. (L.)

Mu-ta' tion Abrupt and heritable modification of a character; also the change ina gene responsible for it. (L. mutates, changed)

Mu' tu-al-ism Jointly beneficial and obligatory association between individuals of two different species. (L. mutuus, exchanged)

My' o-mere A muscle segment or somite. (Gr. mys, muscle + meros, part)

Na' ris; pl. **na'res** The opening of the air passages, both internal and external, in the head of a vertebrate. (L. nostril)

Na' sal Pertaining to the nose.

Nat' u-ral se-lec' tion The elimination of less fit individuals in the struggle to live.

Ne' o-teny Condition of having the period of immaturity indefinitely prolonged as in the axolotl. (Gr. neas, new + teinein, extended)

Ne- phrid' i-um A tubular excretory organ found in mollusks, annelids, arthropod, and other invertebrates. (Gr. nephros, kidney)

Neph' ro-stome The ciliated entrance from the coelomic cavity into a nephridium, or kidney tubule. (Gr. nephros, kidney + stoma, mouth)

Nerve A bundle of nerve fibers lying outside the central nervous system.

Nerve cord A compact cord, composed of neurons and, composed of neurons and usually with gangli, forming part of a central nervous system.

Neu' ral Pertaining to the nervous system. (Gr. neuron, nerve)

Neu' ron A nerve cell with cytoplasmic extensions (dendrites, axon) over which nervous impulses pass.

Noc- tur' nal Active at night. (L. noctumus, nightly)

No'to-chord The elastic cellular axial support formed ventral to the nerve cord in the early embryo of all chordates; later surrounded or supplanted by the vertebrae in most vertebrates. (Gr. notos, the back + chorde, string)

Nu-cle' o-lus An oval mass within the nucleus of most cells, of uncertain function but disappears during mitosis. (L. dim. of nucleus)

Nu' cle- us A differentiated structure of specialized protoplasm within a call refractile and with deeply staining chromatin that controls metabolic activities in cells of all organisms except bacteria. (L. dim. of nux, nut)

O-cel ' lus A small simple eye as on many invertebrates .(L. dim. of oculus, eye)

Oc' u-lar Pertaining to the eye.

Ol-fac' to-ry Pertaining to the sense of smell. (L. olfacto, smell)

Om- niv ' o-rous Eating all kinds of food; feeding upon both plants and animals. (L. omnis, all + voro, eat)

On- tog' e-ny Development of the individual. (Gr. on, being + gen, become)

O-per' cu-lum The plate covering the gills of a bony fish also, the plate serving to cover the opening of some snail shells. (L.)

Oph- thal' mic Pertaining to the eye . (Gr. ophthalmos, eye)

O-pis' tho-coe" lous Concave behind, as the centrum of some vertebrae. (Gr. opisthen, behind + koilos, hollow)

Op' tic Pertaining to the eye or sense of sight. (Gr. optos, seen)

O'ral Pertaining to or near the mouth . (L. os, mouth)

Or'bit The eye socket .(L. orbis, circle)

Or' gan Any part of an animal performing some definite function; a group of cells or tissues acting as a unit for some special purpose . (Gr. organon, instrument)

Or- gan- elle' A specialized part in a protozoan that performs some special function (like an organ in a metazoan). (Gr. dim. of organon, instrument)

Or' gan- ism A single plant or animal; one that functions as a unit.

Os-mo' sis Diffusion through a semipermeable membrane . (Gr. osmos, pushing)

Os'ti – um; pl. os' ti-a An opening to a passage, usually guarded by a valve or circular muscle. (L. os, mouth)

O'tic Pertaining to the ear. (Gr. otikos, pertaining to the ear)

O' to-lith A concretion of calcium salts in the inner ear of vertebrates or in the auditory organ of some in vertebrates. (?Gr. otikos, pertaining to the ear + lithos, stone)

O'vary The organ in which the egg cells are produced and are nourished . (L.ouum, egg)

O' vi-duct The tube by which eggs are conveyed from the ovary to the uterus or to the exterior. (L. ovum, egg + duco, lead)

O-vip' a-rous Egg-laying; Producing eggs that hatch outside the mother's body. (L. ovum, egg + pario, produce)

O' vi-pos" i-tor Organ consisting of paired abdominal appendages variously modified to lay eggs, sometimes slender for boring in wood or sharp and barbed as a sting. (L. ovum, egg + pont, to place)

O' vo- vi- vip" a- rous Producing eggs that are incubated and hatched within the parent's body, as with some fishes, reptiles, and invertebrates. (L. ovum, egg + vivus, alive + pario, produce)

O'vum An egg the sex cell of a female. (L.)

P. The first parental generation __ parents of a given individual of the f1 generation.

Pae" do-gen' e-sis Reproduction by larvae or other preadult forms.(Gr. pais, child + genesis)

Palp (or **palpus**) A projecting part or appendage, often sensory, on the head or near the mouth in some invertebrates. (L. palpo, stroke)

Pa- pil' la Any nipple- like structure, large or small. (L. nipple)

Par' a-site An organism that lives on or in another more or less at the expense of the latter (the host).(Gr. para, beside + sitos, food)

Pa-ren' chy-ma Soft cellular substance filling space between organs. (Gr. para, beside + en, in + chyma, fluid)

Par' the-no-gen" e-sis Development of a new individual from an unfertilized egg as in rotifers, plant lice etc. (Gr. parthenos, virgin + genesis, origin)

Path' o-gen" ic Causing or productive of disease.(Gr. pathos, suffering + genesis, origin)

Pac' to-ral Pertaining to the upper thoracic region, or breast. (L. pectoralis, pertaining to the breast)

Pe- lag 'ic Pertaining to the open sea away from the shore. (Gr. pelagos, open sea)

Pel' vic Pertaining to the posterior girdle and paired appendages of vertebrates; the posterior abdominal redion of a mammal. (L. a basin)

Pe' nis The copulatory organ of a male for conveying sperm to the genital tract of a female.(L.)

Pen' ta-dac"tyl Having five fingers toes, or digits (Gr. pente, five + daktylos, finger)

Per' i-car" di-um The cavity enclosing the heart; also the membranes lining the cavity and covering the heart. (Gr. peri, around + kardia, heart)

Pe- riph' er-al To or toward the surface away from the center. (Gr. pen, around + phero, to bear)

Per' i-stal" sis Rhythmic involuntary muscular contractions passing along a hollow organ, especially of the digestive tract. (Gr. peri, around + stalsis, constriction)

Per' i-to-ne" um The thin serous membrane (mesodermal) that lines the body cavity and covers the organs therein in many animals. (Gr. peri, around + teino, stretch)

Phag' o-cyte A white blood cell that engulfs and digests bacteria and other foreign materials. (Gr. phagein, to eat + kytos, hollow vessel)

Phar' ynx The region of the digestive tract between the mouth cavity and the esophagus; often muscular and sometimes with teeth in invertebrates; the gill region of many aquatic vertebrates. (Gr. throat)

Phe' no-type The external appearance of an individual without regard to its genetic or hereditary constitution compare Genotype. (Gr. phaino, show + typos, impression, type)

Pher' o-mone A chemical signal transmitted between members of the same species. (Gr. pherein, to carry + horman, to excite)

Pho' to- syn" the- sis The formation of carbohydrates from carbon dioxide and water by the chlorophyll in green plants or flagellate protozoans in the presence of light. (Gr. phos, light + synthesis, place together)

Phy-log ' e- ny The history (evolution) of a species or higher group. (Gr. phylon, race + gen, become)

Pig' ment Coloring matter.

Pin' na A wing or fin, also, the projecting part of the external ear in mammals. (L.feather)

Pla- cen' ta The organ by which the fetus (embryo) of higher mammals is attached in the uterus of the mother and through which diffusible substances pass for the metabolism of the fetus. See Chorion. (L. flat cake)

Plan' ti- grade Walking on the whole sole of the foot as a human or bear. (L. planta, sole + gradior, walk)

Plas' ma The fluid portion of blood or lymph. (Gr. athing molded)

Pleu' ra The membrane covering the lung and lining the inner wall of the thorx. (Gr. rib, side)

Pleu' ron The lateral plate on either side of a somite in a arthropods. (Gr. rib, side)

Plex' us A network of interlaced nerves or blood vessels. (L. a plaiting)

Poi' kilo- ther" mal Having varying body temperature; characteristic of all animals but birds and mammals. (Gr. poikilos, variegated + thermal)

Pol' y- mor" phism Existence of individuals of more than one form in a species. (Gr. poly, many + morphe, form)

Pol' yp One of two types of individuals or morphs found in cnidarians. Usually sedentary and reproducing asexually. (Gr. polypous, many footed)

Pol' y-phy- let" ic From more than one known evolutionary derivation. (Gr. poly, many + phyle, tribe)

Pol- y' poid Having three or more times the number of chromosomes.

Pop' u-la" tion The aggregate of individuals of a species that inhabit a particular locality or region.

Por' tal vein One that divides into capillaries before reaching the heart. (L. porta, gate)

Pos- te' ri-or The hinder part or toward the hinder (tail) end, away from the head. Opposite of anterior. (L. following)

Pred' a-tor An animal that captures or preys upon other animals for its food. (L. praeda, booty)

Pre- hen' sile Adapted for grasping or holding. (L. prehendo, seize)

Prim' i- tive Not specialized; the early or beginning type of stage. (L. primus, first)

Pro- coe' lous Concave in front, as the centrum in some vertebrae. ((Gr. pro, before + koilos, hollow)

Proc' to-de" um The ectoderm – lined terminal part of the digestive tract near the anus. (Gr. proktos, anus + hodos, a way)

Pro-sto' mi-um The preoral segment in annelids. (Gr. pro, before + stoma, mouth)

Pro-tan' dry Production of sperm and later of eggs by the same gonad. (Gr. protos, first + aner, man)

Pro'to- ne- phrid" i-um An invertebrate excretory organ (of one or more cells), the inner and closed and either branched or with one terminal cell (solenocyte). See also flme Cell. (Gr. protos, first nephros, kidney)

Pro' to-plasm Living substance; the complex colloidal physicochemical system that constitutes living matter and is the viscid, semifluid material of animal and plant cells.(Gr. protos, first + plasma, form)

Prox' i-mal Relatively nearer the place of attachment or the center of the body. Opposite of distal. (L. proximus, nearest)

Pseu' do- coel A body cavity not lined with peritoneum and not part of a blood –vascular system ; as in nematodes and some other invertebrates. Compare Coelom, Hemocoel. (Gr. pseudes, false + koilos, hollow)

Pseu' do-po" di- um A flowing extension of protoplasm used in locomotion or feeding by a cell or protozoan. (Gr. pseudes, false + podion, foot)

Pul' mo- na- ry Pertaining to the lungs . (L. pulmo,lungs)

Ra' dial sym' me-try Having similar parts arranged around a common central axis, as in a starfish.

Ra' mus A branch or outgrowth of a structure. (L. branch)

Ra' cent The present or Holocene epoch in geology. Compare fossil. (L. recens, fresh)

Re- cep' tor A free nerve ending or sense organ capable of receiving and transforming certain environ mental stimuli into sensory nerve impulses. (L. receiver)

Re- ces' sive char' ac-ter A character from one parent that remains undeveloped in offspring when associated with the corresponding dominant character from the other parent. (L. recessus, a going back)

Rec'tum The terminal enlarged portion of the digestive tract. (L. rectus, straight)

Re- duc' tion di-vi' sion That division of the maturing germ cells by which the somatic or diploid number of chromosomes is reduced to the haploid number. (see also Meiosis)

Re' flex ac' tion Action resulting from an afferent sensory impulse on a nerve center and its reflection as an efferent motor impulse independent of higher nerve centers or the brain. An automatic response to a stimulus. (L. re, back + flecto, bend)

Re- gen' er-a" tion Replacement of parts lost through mutilation or otherwise .

Rre' nal Pertaining to a kidney. (L. renes, kidneys)

Rr' pro-duc' tion The maintenance of a species from generation to generation.

Res' pi- ra" tion Obtaining oxygen from the surrounding medium and giving off carbon dioxide. (L. re, back + spiro, breathe)

Ret' i-na The cell layer of an eye containing the receptors of light impulses .(L. rete, net)

Re-ver' sion The reappearance of ancestral traits that have been in abeyance for one or more generations. (L. re, back + verto, turn)

Ros' trum A projecting snout or similar process on the head. (L. beak)

Ru' di-men" ta-ry Incompletely developed or having no function. Compare Vestigial. (L. rudis, unwrought)

Ru'mi-nant A herbivorous land mammal that chews a cud as a cow or deer.(L. rumen, throat)

Sa' crum The posterior part of the vertebral column that is attached to the pelvic girdle. (L. from scer, sacred, offered in sacrifice)

Sag' it-tal Of or pertaining to the medina anteroposterior plane in a bilaterally symmetrical animal, or a section parallel to that plane. (L. sagitta, arrow)

Sal' i-va-ry Pertaining to the glands of the mouth that secrete saliva. (L. saliva, spittle)

Sap' ro-phyte An organism that lives upon dead organic matter. (Gr. sapros, rotten + phyton, plant)

Scan-so'ri-al Pertaining to or adapted for climbing . (L. scando, climb)

Schi- zog' o-ny Multiple asexual fission in protozoa . (Gr. schizo, to split + gonos, generation, seed)

Sec'on-da-ry sex'u-al char' ac-ters Those characters which distinguish one sex from the other, not functioning directly in reproduction.

Se-cre'tion A useful substance produced in the body by a cell or multicellular gland; also, the process of its production and passage. Compare Ex-cretion. (L. secretus, separated)

Sed'en-ta-ry Remaining in one place .(L. sedeo, sit)

Sed'i-men "ta-ry In geology, rocks formed of calcium carbonates, clay, mud, sand, or gravel, deposited in water or depressions on land and cemented or pressed into solid form. Fossils occur in such rocks. (L. sedimentum, settling)

Seg'ment A part that is marked off or separate from others; any of the several serial divisions of a body or an appendage. Compare Somite.

Sem'i-nal Pertaining to structures or fluid containing spermatozoa (semen). (L. semen, seed)

Sense or'gan An organ containing a part sensitive to a particular kind of stimulus.

Sep 'tum A dividing wall or partition between two cavities or structures .(L. sepes, fence)

Se'rous Secreting watery, colorless serum, as by a gland or serous membrane. (L.)

Se'rum The plasma of blood that separates from a clot and contains on cells or fibrin .(L.)

Ses'sile Permanently fixed, sedentary, not free moving . (L. sedeo, sit)

Se' ta (chaeta) A bristle or slender, stiff, bristle-like structure. (L. bristle)

Sex chro' mo-somes Special chromosomes, different in males and females, and concerned in the determination of sex,the x and y chromosomes.

Sex- lim' i-ted char' ac-ter A character belonging to one sex only , commonly a secondary sexual character.

Sex-linked char' ac-ter A character the gene of which is located in the sex chromosome.

Sex' u-al u' nion Temporary connection of a male and female for transfer of sperm into the female's reproductive tract.

Si-li' ceous Containing silica or silicon dioxide (SiO_2).(L. silex, flint)

Si' nus A cavity in a bone or an enlargement in a blood vessel. (L. fold, hollow)

Skel' e-ton The hardened framework of an animal body serving for support and to protect the soft parts, it may be external or internal and either solid or jointed. (Gr.)

Sol' I ta-ry Living alone; not in colonies or groups.(L. solus, alone)

Sol'ute A substance that will dissolve or go into solution, as salt in water . (L. se, apart + lus, set free)

Sol' vent A fluid capable of dissolving substances.

So'ma, so-mat' ic Pertaining to the body or body cell, as contrasted with germ cells. (Gr. soma, body)

So' mite A serial segment or homologous part of the body; a metamere. (Gr. soma, body)

Spe' cial- ized Not primitive; adapted by structure or by function for a particular purpose or mode of life.

Spe' cies (pl. also **species**) The unit in classification of animals or plants. (see Chap. 14.)

Sperm See spermatozoa.

Sper' ma- to- phore A packet of sperm extruded by the male and transferred to the female. (Gr. sperma, seed + phero, to bear)

Sper' ma- to- zo" a The matured and functional male sex cells or male gametes. (Gr. sperma, seed + zoom, animal)

Spir' a- cle In insects, an external opening to tracheal or respiratory system; in cartilaginous fishes, the modified first gill slit. (L. spiraculum, air hole)

Spore A cell in a resistant covering capable of developing independently into a new individual. (Gr. spora, seed)

Stat' o-cyst An organ of equilibrium in some invertebrates.(Gr . statos, standing + kystis, bladder)

Stat' o-lith A calcareous granule in a ststocyst.

Stim' u-lus A change in the external or internal environment capable of influencing some activity in an organism or its oarts.

Sto' mo-de" um The ectoderm-lined portion of the mouth cavity . (Gr. stoma, mouth + daio, divide)

Strat' i- fied A series of layers, one above another. (L.)

Stra' tum A layer or sheet of tissue (anatomy); a layer or sheet of sedimentary rock (geology). (L. covering)

Su' ture Line of junction between two bones or between two parts of an exoskeleton . (L. suo, sew)

Sym' bi-o"sis Interrelation between two organisms of different species, see Mutualism, parasite. (Gr. syn, togther + bios, life)

Sym' phy- sis A union between two parts. (Gr. syn, together + phyein, grow)

Syn- apse' The contact of one nerve cell with another, across witch impulses are transmitted.(Gr. syn, together + hapto, unite)

Syn- ap' sis Temporary union of the chromosomes in pairs preliminary to the first maturation division. (Gr. syn, together + hapto, unite)

Syn- cyt' i-um A mass or layer of protoplasm containing several or many nuclei not separated by cell membranes. (Gr. syn, together + kytos, cell)

Sys- tem' ic Portion of the circulatory system not directly involved in respiration.

Tac' tile Pertaining to the organs of the sense of touch. (L. tactus, touch)

Tec' tin The organic material in skeletons of some protozoans; pseudochitin.

Ten' don A connective tissue band attaching a muscle. (L. tendo stretch)

Ten' ta-cle An elongate flexible appendage usually near the mouth . (L. tento, from teneo, hold)

Ter- res' tri-al Belonging to or living on the ground or earth. (L. treea, earth)

Tes' tis;pl. **tes' tes** The male germ gland or gonad, in which spermatozoa are formed. (L.)

Tet' ra-pod A vertebrate typically with four limbs__ the amphibians, reptiles, birds, and mammals. (Gr. tetra, four + pous, foot)

Tho' rax The major division of an animal next behind the head; in land vertebrates, the part enclosed by the ribs. (Gr.)

Tis' sue A layer or group of cells in an organ or body part having essentially the same structure and function.(L. texo, weave)

Tra'che-a An air tube; the windpipe of land vertebrates from the glottis to the lungs; part of the respiratory system in insects and other arthropods. (Gr. trachys, rough)

Trip' lo-blas" tic Derived from three embryonic germ layers__ ectoderm, endoderm, and mesoderm. (Gr. triplous, threefold + blastos, germ)

Tro' cho- phore An invertebrate larva, commonly top-shaped and with an equatorial band of cilia. (Gr. trochos, wheel + phoros, bearing)

Tym' pa-num A vibrating membrane involved in hearing; the eardrum, or tympanic membrane. (Gr. tympanon, drum)

Um-bil' i-cal cord The cord containing blood vessels supported by connective tissue that unites the embryo or fetus of a mammal with the mother during development in the uterus. (L. umbilicus, navel)

Un-guic' u-late Having claws, as a cat. (L. unguis, claw)

Un' gu-late Having hoofs, as a deer or horse. (L.ungula, hoof)

Un' gu-l- grade' Walking or adapted for walking on hoofs. (L. ungula, hoof + gradior, walk)

Unit character A trait that behaves more or less as a unit in heredity, and may be inherited independently of other traits.

U-re' ter The duct carrying urine from the kidney to the urinary bladder or to the cloaca. (Gr. from ouron, urine)

U-re' thra The duct by which urine is discharged from the bladder to the outside in mammals, joined by the vasa deferentia in the male. (Gr. from ouron, urine)

U' ri- no- gen" i-tal See Urogenital.

U' ro-gen " i-tal Pertaining to the excretory and reproductive organs and functions .(L. urina, urine + genitalis, genital)

U' ter-us The enlarged posterior portion of an oviduct in which eggs may be retained for development. (L. womb)

Vac' u-ole A minute cavity within a cell, usually filled with some liquid product of protoplasmic activity.(L. dim. of vacuus, empty)

Va- gi' na The terminal portion of the female reproductive tract, which receives the ulatory organ of a male in mating.(L. sheath)

Valve In animals, any structure that limits or closes an opening the thin folds in veins, lymph vessels, or hearts or the circular muscles about a tubular exit; also, either external shell of a bivalve mollusk, brachiopod, or some crustaceans.

Vas; pl.**va' sa** A small, tubular vessel, or duct, especially one leading from the testis. (L. vessel)

Vas' cu- lar Pertaining to vessels or ducts for conveying or circulating blood or lymph. (L. dim. of vas, vessel)

Vein A vessel carrying blood from capillaries toward the heart. (L. vena, vein)

Vent The external opening of the cloaca or the intestine on the surface of the body, especially in non mammals, as birds, fish, and reptiles.

Ven' tral Toward the lower side or belly; away from the back .Opposite to Dorsal.(L. venter , belly)

Ven' tri- cle The muscular chamber in a heart; also, a cavity in the brain of a vertebrate. (L. dim. of venter, belly)

Ver' te- bra One of the segmental structural units of the axial skeleton or spinal column in a vertebrate. (L. joint)

Ver' te- brate An animal having a segmental "backbone" or vertebral column; the cyclostomes to mammals. (L. vertebrates, jointed)

Ves' sel A tubular structure that conveys fluid, especially blood or lymph . (L. vascellum , dim. of vas, vessel)

Ves- tig' i-al Small or degenerate but representing a structure that formerly was more fully developed or functional. Compare rudimentary. (L. vestigium, footstep, a trace)

Vil' lus; pl.vil' li A minute, finger- like projection; especially those numerous on the intestinal lining of vertebrates. (L. villus, shaggy hair)

Vis' cer-a The organs within the cranium, thorax, and abdomen, specially the latter. (L. internal organs)

Vis' cer-al skel' e-ton The supporting framework of the jaws and gill arches and their derivatives, in vertebrates.

Vis' u-al Pertaining to sight.

Vi'ta- min An organic substance that is an essential food factor needed in minute amounts for normal growth and function. (L. vita, lite + amine, a chemical radical)

Vi-vip' a-rous Producing living young that develop from eggs retained within the mother's body and nourished by her bloodstream, as with most mammals. (L. vivo, live + pario, to bear)

X,Y chromosomes Chromosomes associated with sex in many animals.

Yolk Fat or oil droplets stored within or with an egg for nourishment of the future embryo.

Zy' gote A fertilized egg resulting from the union of two gametes of opposite kind ovum and sperm. (Gr. zygon, yok)

www.ingramcontent.com/pod-product-compliance
Lightning Source LLC
Chambersburg PA
CBHW042337150426
43195CB00001B/23